Selected Titles in This Series

730 **Suhyoung Choi,** The decomposition and classification of radiant affine 3-manifolds, 2001

729 **Michael Grosser, Eva Farkas, Michael Kunzinger, and Roland Steinbauer,** On the foundations of nonlinear generalized functions I and II, 2001

728 **Laura Smithies,** Equivariant analytic localization of group representations, 2001

727 **Anthony D. Blaom,** A geometric setting for Hamiltonian perturbation theory, 2001

726 **Victor L. Shapiro,** Singular quasilinearity and higher eigenvalues, 2001

725 **Jean-Pierre Rosay and Edgar Lee Stout,** Strong boundary values, analytic functionals, and nonlinear Paley-Wiener theory, 2001

724 **Lisa Carbone,** Non-uniform lattices on uniform trees, 2001

723 **Deborah M. King and John B. Strantzen,** Maximum entropy of cycles of even period, 2001

722 **Hernán Cendra, Jerrold E. Marsden, and Tudor S. Ratiu,** Lagrangian reduction by stages, 2001

721 **Ingrid C. Bauer,** Surfaces with $K^2 = 7$ and $p_g = 4$, 2001

720 **Palle E. T. Jorgensen,** Ruelle operators: Functions which are harmonic with respect to a transfer operator, 2001

719 **Steve Hofmann and John L. Lewis,** The Dirichlet problem for parabolic operators with singular drift terms, 2001

718 **Bernhard Lani-Wayda,** Wandering solutions of delay equations with sine-like feedback, 2001

717 **Ron Brown,** Frobenius groups and classical maximal orders, 2001

716 **John H. Palmieri,** Stable homotopy over the Steenrod algebra, 2001

715 **W. N. Everitt and L. Markus,** Multi-interval linear ordinary boundary value problems and complex symplectic algebra, 2001

714 **Earl Berkson, Jean Bourgain, and Aleksander Pełczynski,** Canonical Sobolev projections of weak type $(1, 1)$, 2001

713 **Dorina Mitrea, Marius Mitrea, and Michael Taylor,** Layer potentials, the Hodge Laplacian, and global boundary problems in nonsmooth Riemannian manifolds, 2001

712 **Raúl E. Curto and Woo Young Lee,** Joint hyponormality of Toeplitz pairs, 2001

711 **V. G. Kac, C. Martinez, and E. Zelmanov,** Graded simple Jordan superalgebras of growth one, 2001

710 **Brian Marcus and Selim Tuncel,** Resolving Markov chains onto Bernoulli shifts via positive polynomials, 2001

709 **B. V. Rajarama Bhat,** Cocylces of CCR flows, 2001

708 **William M. Kantor and Ákos Seress,** Black box classical groups, 2001

707 **Henning Krause,** The spectrum of a module category, 2001

706 **Jonathan Brundan, Richard Dipper, and Alexander Kleshchev,** Quantum Linear groups and representations of $GL_n(\mathbb{F}_q)$, 2001

705 **I. Moerdijk and J. J. C. Vermeulen,** Proper maps of toposes, 2000

704 **Jeff Hooper, Victor Snaith, and Min van Tran,** The second Chinburg conjecture for quaternion fields, 2000

703 **Erik Guentner, Nigel Higson, and Jody Trout,** Equivariant E-theory for C^*-algebras, 2000

702 **Ilijas Farah,** Analytic guotients: Theory of liftings for quotients over analytic ideals on the integers, 2000

701 **Paul Selick and Jie Wu,** On natural coalgebra decompositions of tensor algebras and loop suspensions, 2000

700 **Vicente Cortés,** A new construction of homogeneous quaternionic manifolds and related geometric structures, 2000

(*Continued in the back of this publication*)

The Decomposition and Classification of Radiant Affine 3-Manifolds

MEMOIRS
of the
American Mathematical Society

Number 730

The Decomposition and Classification of Radiant Affine 3-Manifolds

Suhyoung Choi

November 2001 • Volume 154 • Number 730 (first of 5 numbers) • ISSN 0065-9266

American Mathematical Society
Providence, Rhode Island

2000 *Mathematics Subject Classification.*
Primary 57M50; Secondary 53A20, 53C15, 53C12.

Library of Congress Cataloging-in-Publication Data

Choi, Suhyoung.
The decomposition and classification of radiant affine 3-manifolds / Suhyoung Choi.
p. cm. — (Memoirs of the American Mathematical Society, ISSN 0065-9266 ; no. 730)
"Volume 154, number 730 (first of 5 numbers)."
Includes bibliographical references.
ISBN 0-8218-2704-9 (alk. paper)
1. Three-manifolds (Topology) I. Title. II. Series.

QA3.A57 no. 730
[QA613.2]
510 s—dc21
[514′.3] 2001034315

Memoirs of the American Mathematical Society

This journal is devoted entirely to research in pure and applied mathematics.

Subscription information. The 2001 subscription begins with volume 149 and consists of six mailings, each containing one or more numbers. Subscription prices for 2001 are $494 list, $395 institutional member. A late charge of 10% of the subscription price will be imposed on orders received from nonmembers after January 1 of the subscription year. Subscribers outside the United States and India must pay a postage surcharge of $31; subscribers in India must pay a postage surcharge of $43. Expedited delivery to destinations in North America $35; elsewhere $130. Each number may be ordered separately; *please specify number* when ordering an individual number. For prices and titles of recently released numbers, see the New Publications sections of the *Notices of the American Mathematical Society.*

Back number information. For back issues see the *AMS Catalog of Publications.*

Subscriptions and orders should be addressed to the American Mathematical Society, P. O. Box 845904, Boston, MA 02284-5904. *All orders must be accompanied by payment.* Other correspondence should be addressed to Box 6248, Providence, RI 02940-6248.

Memoirs of the American Mathematical Society is published bimonthly (each volume consisting usually of more than one number) by the American Mathematical Society at 201 Charles Street, Providence, RI 02904-2294. Periodicals postage paid at Providence, RI. Postmaster: Send address changes to Memoirs, American Mathematical Society, P. O. Box 6248, Providence, RI 02940-6248.

This publication is indexed in *Science Citation Index*®, *SciSearch*®, *Research Alert*®, *CompuMath Citation Index*®, *Current Contents*®/*Physical, Chemical & Earth Sciences.*
Printed in the United States of America.

♾ The paper used in this book is acid-free and falls within the guidelines established to ensure permanence and durability.
Visit the AMS home page at URL: `http://www.ams.org/`

10 9 8 7 6 5 4 3 2 1 06 05 04 03 02 01

Contents

Abstract

An affine manifold is a manifold with torsion-free flat affine connection. A geometric topologist's definition of an affine manifold is a manifold with an atlas of charts to the affine space with affine transition functions; a radiant affine manifold is an affine manifold with a holonomy group consisting of affine transformations fixing a common fixed point. We decompose a closed radiant affine 3-manifold into radiant 2-convex affine manifolds and radiant concave affine 3-manifolds along mutually disjoint totally geodesic tori or Klein bottles using the convex and concave decomposition of real projective n-manifolds developed earlier. Then we decompose a 2-convex radiant affine manifold into convex radiant affine manifolds and concave-cone affine manifolds. To do this, we will obtain certain nice geometric objects in the Kuiper completion of a holonomy cover. The equivariance and local finiteness property of the collection of such objects will show that their union covers a compact submanifold of codimension zero, the complement of which is convex. Finally, using the results of Barbot, we will show that a closed radiant affine 3-manifold admits a total cross-section, confirming a conjecture of Carrière, and hence every closed radiant affine 3-manifold is homeomorphic to a Seifert fibered space with trivial Euler number, or a virtual bundle over a circle with fiber homeomorphic to a Euler characteristic zero surface. In Appendix C, Thierry Barbot and the author show the nonexistence of certain radiant affine 3-manifolds and that compact radiant affine 3-manifolds with nonempty totally geodesic boundary admit total cross-sections, which are key results for the main part of the paper.

Received by the editor July 19, 2000.

Appendix C: "Radiant affine 3-manifolds with boundary, and certain radiant affine 3-manifolds," written by Thierry Barbot (UMPA, École Normale Supérieure de Lyon, 46, alée d'Italie, LYON, France) and Suhyoung Choi.

Key words and phrases. real projective structure, affine 3-manifold, affine structure, geometric structure, flat connection, flow, foliation.

S. Choi's research partially supported by GARC-KOSEF, the Ministry of Education 1997-001-D00036, and the BK21 program of the Ministry of Education. T. Barbot's research was supported by the CNRS.

A part of this result was announced at the conference "Combinatorial problems arising in knots and 3-manifolds" January 21-24, 1997 MSRI, Berkeley, CA.

CHAPTER 0

Introduction

The classical (X, G)-geometry is the study of invariant properties on a space X under the action of a Lie group G as Felix Klein proposed. An (X, G)-structure on a manifold M prescribes local identification of M with X with transition functions in G. An (X, G)-structure is also given by an immersion, so-called developing map, from the universal cover of the manifold M to X equivariant with respect to a homomorphism, so-called holonomy homomorphism, from the deck-transformation group $\pi_1(M)$ to G. Differential geometers can define (X, G)-structure using higher order curvatures.

Two of the fundamental general questions of geometric topologists are to find a topological obstruction to the existence of structures locally modeled on (X, G) and classify such structures on a given manifold up to (X, G)-self-diffeomorphisms.

There are many important (X, G)-structures. In particular, when X has a G-invariant complete metric, then the study is equivalent to the study of discrete cocompact subgroups of Lie groups which is well-established and gave us many fruitful topological implications.

We will be working in real projective and affine structures on manifolds, a very much open area presently. Recall that real projective geometry is given by a pair $(\mathbf{R}P^n, \mathrm{PGL}(n+1, \mathbf{R}))$ where $\mathrm{PGL}(n+1, \mathbf{R})$ is the projectivized general linear group acting on $\mathbf{R}P^n$. An affine geometry is given by a pair $(\mathbf{R}^n, \mathrm{Aff}(\mathbf{R}^n))$ where $\mathrm{Aff}(\mathbf{R}^n)$ is the group of affine transformations of $\mathbf{R}^n$; i.e., transformations of form $x \mapsto Ax+b$ for A an element of the general linear group $\mathrm{GL}(n, \mathbf{R})$ and b an n-vector. A *real projective n-manifold* is an n-manifold with a geometric structure modeled on real projective geometry; an *affine n-manifold* one with that modeled on affine geometry. (In differential geometry, an affine manifold is defined as a manifold with a flat torsion-free affine connection. A real projective n-manifold is equivalent to a manifold with a projectively flat torsion-free affine connection.)

An important distinguishing feature of these geometric manifolds is that geodesics are defined by connections. Basically, they correspond to usual straight lines under charts.

A sphere with real projective structure induced from the standard double covering map is said to be a *real projective sphere*. A compact real projective disk with geodesic boundary real projectively diffeomorphic to a closed hemisphere in $\mathbf{S}^2$ is said to be a *real projective hemisphere*. These are trivially unique structures.

Affine and real projective structures on annuli or tori were classified by Nagano-Yagi [**38**] and Goldman [**21**]. Benzécri [**7**] showed that any affine surface must be homeomorphic to tori or annuli.

A real projective manifold is *convex* if its universal cover is real projectively homeomorphic to a convex domain in an affine patch of $\mathbf{R}P^n$. They form the most important class of real projective manifolds. As hyperbolic geometry is a geometry included in projective geometry by the Klein model, we see that every hyperbolic n-manifold carry a canonical real projective structure. Also, there are nontrivial deformations of these real projective manifolds to convex real projective manifolds as shown by Kac-Vinberg [**28**], Koszul [**36**] and Goldman [**23**].

Many examples of affine manifolds were constructed by Sullivan-Thurston [**44**] from their examples of complex projective or Möbius structures on surfaces and manifolds.

The *holonomy group* is the image of the holonomy homomorphism. The purpose of this paper is to show that the above two questions can be answered for radiant affine structures on 3-manifolds. A so-called radiant affine manifold is an affine manifold whose holonomy group fix a common point of $\mathbf{R}^n$; i.e., an atlas can be chosen so that transition functions are in the subgroup $\mathrm{GL}(n, \mathbf{R})$ of $\mathrm{Aff}(\mathbf{R}^n)$. One can think of a radiant affine manifold as a manifold with $(\mathbf{R}^n, \mathrm{GL}(n, \mathbf{R}))$-structure, where $\mathrm{GL}(n, \mathbf{R})$ acts on $\mathbf{R}^n$ in the standard manner.

We recall that examples of radiant affine $(n+1)$-manifolds can be obtained by so-called generalized affine suspensions over real projective n-manifolds: We realize a real projective manifold times a real line as an affine manifold of dimension $n+1$ by adding the radial direction in a natural manner. Then we take a quotient by an infinite cyclic group acting affinely and mostly in the radial direction to make a compact affine quotient manifold. A generalized affine suspension is said to be a *Benzécri suspension* if the group acts purely in the radial direction up to finite order (see Chapter 3 and Appendix C for more details).

A submanifold of a real projective or affine manifold is *totally geodesic* if each point of it has a neighborhood with a chart mapping the manifold into a subspace of the real projective space or the affine space respectively. Given a real projective or affine manifold M, let N be a totally geodesic $(n-1)$-dimensional manifold properly imbedded in the interior M^o of M. Then we define the splitting of M along N as taking the disjoint unions of the completions of all components of $M - N$ (see Chapter 1 or [**15**] for more details). We say that a real projective or affine manifold M *decomposes* along N to another real projective or affine manifold M' if M' is obtained from M by splitting along N. In this paper, we will prove the following result:

THEOREM A . *Let M be a compact radiant affine 3-manifold with empty or totally geodesic boundary. Then M decomposes along the union of finitely many disjoint totally geodesic tori or Klein bottles, tangent to the radial flow, into*

1. *convex radiant affine 3-manifolds*
2. *generalized affine suspensions of real projective spheres, real projective planes, real projective hemispheres, or π-annuli (or Möbius bands) of type C; or affine tori, affine Klein bottles, or affine annuli (or Möbius bands) with geodesic boundary.*

Note this decomposition is not canonical; however, the statements can be made into ones about canonical decomposition with additional requirements. A *π-annulus* of type C is the third type of a π-annulus in Section 3 of [**12**], i.e., a π-annulus that is the sum of two elementary annuli of type IIb and with nondiagonalizable holonomy for the generator of the fundamental group. A *π-Möbius band of type C* is a real projective Möbius band that is doubly covered by a π-annulus of type C.

The radiant vector field on $\mathbf{R}^n$ as given by $\sum_{i=1}^n x_i \partial/\partial x_i$ is $\mathrm{GL}(n, \mathbf{R})$-invariant, and hence induces a vector field and a flow on M by $\mathbf{dev}$ and the covering map.

The vector field is said to be a *radiant vector field*, and the flow the *radial flow*. A *total cross-section* is a closed transverse submanifold that meets every flow line. Carrière [**10**] asked whether every compact radiant affine 3-manifold admits a total cross-section to the radial flow. In the paper, Carrière showed that if the radial flow preserves a volume form, then this is true. In dimension 6, Fried produced a counter-example [**19**].

Barbot has shown that the conjecture is true if the holonomy group is virtually solvable (and more generally, if some finite-index subgroup preserves a plane of $\mathbf{R}^3$), or if the manifold is homeomorphic to a Seifert 3-manifold (see [**2**] and [**4**]). Barbot confirms the conjecture if the 3-manifold has a totally geodesic surface tangent to the radial flow, if there exists a closed orbit of non-saddle type, or if the 3-manifold is convex (see Theorem 3.3 and [**3**]). Barbot and Choi in Appendix C, i.e., Theorem C.3, show that if the boundary is totally geodesic and nonempty, then the 3-manifold has to be an affine suspension. These results enable us to answer the Carrière conjecture in the positive:

COROLLARY A . *Let M be a compact radiant affine 3-manifold with empty or totally geodesic boundary. Then M admits a total cross-section to the radial flow. As a consequence M is affinely diffeomorphic to one of the following affine manifolds*:

- *a Benzécri suspension over a real projective surface of negative Euler characteristic with empty or geodesic boundary,*
- *a generalized affine suspension over a real projective sphere, a real projective plane, or a hemisphere,*
- *a generalized affine suspension over a π-annulus (or Möbius band) of type C; or an affine torus, an affine Klein bottle, an affine annulus (or Möbius band) with geodesic boundary.*

The classification of Benzécri suspensions and generalized affine suspensions over real projective surfaces of zero Euler characteristic reduces to the classification of real projective orbifolds of negative Euler characteristic and the classification of real projective automorphisms of real projective tori or Klein bottles (see [**2**]).

We obtain the following topological characterization of radiant affine 3-manifolds. (See Scott [**40**] for definition of a Euler number of a Seifert fibration.)

COROLLARY B . *Let M be a compact radiant affine 3-manifold with empty or totally geodesic boundary. Then M is homeomorphic to a Seifert space of Euler number 0 or a virtual bundle over a circle with fiber homeomorphic to a Euler characteristic zero compact surface.*

The purpose of this paper is to prove Theorem A, approaching by geometric techniques developed initially for two-dimensions in [**11**] and [**12**] and later generalized to n-dimensions in the monograph [**15**].

In Chapters 1, 2, 3 and Appendices A and B, we will discuss general n-dimensional real projective and affine manifolds while there are no losses from studying in the general dimensions. The main arguments for Theorem A are carried out in Chapters 4-10: First, using the decomposition result obtained in [**15**], we decompose a radiant affine manifold into 2-convex submanifolds and radiant concave affine submanifolds (see Definition 2.2 in Chapter 2). A real projective 3-manifold is 2-*convex* if a nondegenerate real projective map from $T^o \cup F_2 \cup F_3 \cup F_4$ for an affine 3-simplex T in $\mathbf{R}^3$ with sides $F_1, \dots, F_4$ always extends to one from

T. A *concave affine* manifold is a real projective manifold with a very special affine structure so that any point of its universal cover is covered by domains affinely homeomorphic to affine half-spaces. Next, we decompose a 2-convex radiant affine manifold into convex affine manifolds and concave-cone affine manifolds (see Definition 10.3 in Chapter 10) in Chapters 6-10. A concave cone affine 3-manifold is a real projective manifold with a special affine structure so that its universal cover is covered by domains affinely homeomorphic to affine quarter spaces. We will prove Theorem A in the final part of Chapter 10 using results from Chapters 11 and 12. In Chapter 11, we show that radiant concave affine manifolds and concave-cone affine manifolds are generalized affine suspensions. In Chapter 12, we discuss a step needed in Chapters 6 and 12. In Appendix C, we show the nonexistence of affine manifolds whose developing maps are universal covering maps of $\mathbf{R}^3$ with a line removed. More importantly, we show that the Carrière conjecture holds if the affine manifold has a nonempty totally geodesic boundary of sufficiently general type; i.e., they admit total cross-sections.

In Chapter 1, we give definitions of real projective structures, developing maps, holonomy covers, and so on. We discuss lifts of the developing map $\tilde{M} \to \mathbf{R}P^n$ to $\mathbf{S}^n$, and the holonomy homomorphism to the group $\mathrm{Aut}(\mathbf{S}^n)$ of projective automorphisms of $\mathbf{S}^n$. A *holonomy covering* M_h of M is the cover of M corresponding to the kernel of a holonomy homomorphism h. M_h is well-defined since h can change only by a conjugation by a real projective automorphism. Then **dev** induces a metric on M_h from the standard Riemannian metric on $\mathbf{S}^n$, and the completing M_h for the metric, we obtain the Kuiper completion $\check{M}_h$ of M_h. We discuss the ideal set or frontier of M_h in $\check{M}_h$, convex subsets in the Kuiper completions, and extending maps from one ball to another one. This type of completion was first introduced by Kuiper [**37**] and is the most useful technical device in studying the incompleteness in geometric structures. (Since this is the most important technical machinery needed in this paper, we advise our readers to become completely comfortable with involved notions.) The ideal set can be thought of as points infinitely far away from points of M_h if M_h is endowed with a Riemannian metric d_M induced from that of the compact manifold M. We also define various polyhedra in $\mathbf{S}^n$ and hence ones on $\check{M}_h$ as well. In particular, a *bihedron* is an n-ball in $\mathbf{S}^n$ that is the closure of a component of the complement of two distinct great $(n-1)$-spheres in $\mathbf{S}^n$. A *trihedron* is one that is the closure of a component of the complement of three great $(n-1)$-spheres in general position. A *tetrahedron* is one for four great $(n-1)$-spheres in general position. Corresponding objects in $\check{M}_h$ are also defined. (This section can be browsed through if the reader is familiar with the field.)

In Chapter 2, we discuss $(n-1)$-convexity, and the decomposition of real projective n-manifolds into $(n-1)$-convex submanifolds and concave affine n-manifolds. However, the submanifold along which we decompose may not be totally geodesic.

In Chapter 3, radiant affine manifolds, regarded as real projective manifolds, are discussed. We define the term generalized affine suspensions of a real projective surface, and show that a generalized affine suspension over a negative Euler characteristic real projective surface is a Benzécri suspension over a real projective surface. We discuss how the radial flow extends to the projective completions, and define the origin, radiant sets, finitely ideal sets, infinitely ideal sets, and discuss the relationship between radiant n-bihedra and n-crescents. A so-called n-crescent is essentially a convex n-bihedron one of whose sides is in the ideal set.

In Chapter 4, we apply materials of Chapters 1, 2, and 3 to three-dimensional radiant affine manifolds. We show that a radiant affine 3-manifold decomposes into 2-convex radiant affine 3-manifolds and radiant concave affine 3-manifolds along mutually disjoint union of totally geodesic tori or Klein bottles. It is proved in Chapter 5 that the decomposing surface is totally geodesic unlike the ones in the general real projective manifolds.

In Chapters 6-10, we study exclusively 2-convex radiant affine 3-manifolds with empty or totally geodesic boundary. We assume that M is not convex in these chapters. The so-called crescent-cone is a radiant trihedron with three sides (which are lunes) so that two of its sides are in the ideal set respectively (see Definition 6.1 and Figure 6.1). We will prove Theorem A and Corollary A, following the basic strategy analogous to that of papers [**11**] and [**15**]:

(1) We find a radiant tetrahedron detecting nonconvexity in $\check{M}_h$ for a compact radiant affine manifold M.
(2) By taking a suitable subsequence of deck transformations, we replace it by a crescent-cone in $\check{M}_h$ (by a blowing up argument as in [**11**]).
(3) Using the equivariance and local finiteness properties of the collection of crescent-cones, we obtain a disjoint collection of concave-cone affine submanifolds in M of codimension 0 with totally geodesic boundary.
(4) We obtain the decomposition: the closures of components of the complement in M of the union consist of convex radiant affine 3-manifolds.

More details are given below:

(a) Since M is not convex, we obtain a triangle in M_h detecting the nonconvexity of M.
(b) Obtain a radiant tetrahedron F by radially extending the triangle. The tetrahedron has four sides F_1, F_2, F_3, and F_4, where F_4 is a subset of infinitely ideal set, and F_1^o, F_2^o those of M_h, and F_3 meets the finitely ideal set and M_h—(1).
(c) We choose an appropriate sequence of points $p_i \in F_3 \cap M_h$ leaving every compact subset of M_h. Choosing a fixed fundamental domain, a sequence of equivalent points q_i in it, and the deck transformation ϑ^i so that $\vartheta^i(p_i) = q_i$, we obtain a sequence of objects in F pulled along with p_i by ϑ^i. We show that $\mathbf{dev}(\vartheta^i(F))$ converges to a radiant 3-ball in $\mathbf{S}^3$. Hence, there exists a radiant convex 3-ball F^u in $\check{M}_h$ to which F^i "converges" (see Appendix B). The convex 3-ball F^u is either a radiant tetrahedron, or a radiant trihedron as F^i are radiant tetrahedra. Using the following claim, we show that there exists a so-called pseudo-crescent-cone or crescent-cone, which is a radiant tetrahedron or a radiant trihedron with exactly one side intersecting M_h and the remaining sides in $M_{h\infty}$. By Chapter 12, we rule out pseudo-crescent-cones—(2).
(d) **Claim**: *We can choose an appropriate sequence of points $p_i \in F_3 \cap M_h$ leaving every compact subset of M_h such that the sequence of the d_M-distances from p_i to $(F_1 \cup F_2) \cap M_h$ goes to infinity* (Proposition 7.1):
To prove the claim, we suppose that for every sequence of points p_i, the d_M-distance is bounded above, and choose p_i in a certain manner. We show this implies contradictions in Chapters 8 and 9 when F^u is a radiant tetrahedron and a radiant trihedron respectively.

(e) In Chapter 10, using the equivariant and locally finite collection of crescent-cones, we obtain a radiant set which covers a concave-cone affine 3-manifold in M —(3). From the classification of radiant concave affine manifolds and concave-cone affine manifolds in Chapter 11, we complete the proof of Theorem A. —(4).

In Chapter 11, we classify radiant concave affine manifolds and concave-cone affine manifolds. We show that they are generalized affine suspensions of Euler characteristic nonnegative real projective surfaces with geodesic boundary. The method uses the work by Tischler and heavily that by Barbot and Choi in Appendix C.

In Chapter 12, we show that pseudo-crescent-cones do not occur if M is not convex.

The materials in Appendices A and B are adopted from the paper [**11**] to general n-dimension. The proofs are identical. (We include these for reader's convenience.) In Appendix A, we discuss dipping intersection of two balls, and transversal intersection of two n-crescents. In Appendix B, we discuss sequences of compact convex n-ball in the Kuiper completion, and discuss their convergence property when they include a common open ball. (See [**15**] for details.)

In Appendix C, written by Barbot and Choi, we show the nonexistence of a radiant affine 3-manifold whose developing map is a universal covering map to $\mathbf{R}^3$ with a line removed. The proof follows from the fact that such a manifold has a total cross-section, which uses the work of Barbot by showing that the holonomy group of such a manifold is either solvable or has a hyperbolic element. The second part of Appendix C is devoted to showing that a compact radiant affine manifold with nonempty boundary has a total cross-section. The boundary here is assumed to be totally geodesic and convex or must be a two-dimensional Hopf manifold. We do this by showing that a certain connected abelian Lie group of rank 2 acts on the manifold giving us a foliation by tori or a decomposition of the manifold into pieces which admit total cross-sections. Again, the theory of Barbot developed for the case when the radiant affine manifold contains a totally geodesic surface tangent to the radial flow is used. These are key steps for the main part of the paper.

Acknowledgement

The author would like to thank Boris Apanasov, Thierry Barbot, Yves Carrière, William Goldman, Yoshinobu Kamishima, Hyuk Kim, Sadayoshi Kojima, Shigenori Matsumoto, William Thurston, and Abdelghani Zeghib for many helpful discussions. The author particularly enjoyed the cooperation with Thierry Barbot generously explaining and supplying him with short arguments, in many occasions enabling the author to complete this work. The author also thank him for writing an appendix with him, providing key results in the boundary case using his approach essentially. Furthermore, the author likes to thank Yves Carrière and David Fried for posing many extremely interesting conjectures in this field. As usual, we need more well-posed problems in this field. With their guidance, the field has become more mature and fertile. Also, the author thanks Sang-Eun Lee for much help on my computer running Linux. The figures were drawn by xfig and Maple. The final portion of this work was completed during the author's visit to the IHES and the ENS-Lyon. The author appreciates the hospitality given by both institutes with great tradition of excellence in geometry of all kinds. The author also thanks

GARC for generous financial support. Finally, the author is very thankful for the support of Haeyon while the author spent numerous hours writing and revising this paper.

CHAPTER 1

Preliminary

In this chapter, we define developing maps and holonomy and discuss the natural relation between affine manifolds and real projective manifolds, some history on the subject of affine and real projective manifolds, developing maps and holonomy homomorphisms lifted to $\mathbf{S}^n$ and the group $\mathrm{Aut}(\mathbf{S}^n)$ of projective automorphisms, the completion $\check{M}$ of the universal cover $\tilde{M}$ of a real projective manifold M due to Kuiper, and the holonomy cover M_h of M. We define convex sets in $\mathbf{S}^n$ and $\check{M}$ and the Kuiper completion $\check{M}_h$ of M_h. We introduce an important lemma that how two balls in $\check{M}_h$ meet may be completely read from their images under the developing map **dev** of M under very natural circumstances.

An (X, G)-structure on a manifold M is given by a maximal atlas of charts to X where the transition functions are in G. An (X, G)-map is a local diffeomorphism preserving (X, G)-structures locally. (Excellent treatments can be found in Ratcliffe [**39**] and Thurston [**45**].) (This will be our view point of (X, G)-structures in this paper. However, there are differential-geometry version of (X, G)-structures which we will not use.)

Given an (X, G)-structure on M, we can associate to it an immersion $\mathbf{dev} : \tilde{M} \to X$, called *a developing map*, and a homomorphism h from the group $\pi_1(M)$ of deck transformations to G, called a *holonomy homomorphism* satisfying $h(\gamma) \circ \mathbf{dev} = \mathbf{dev} \circ \gamma$ for each $\gamma \in \pi_1(M)$. **dev** is obtained by analytically extending coordinate charts of $\tilde{M}$ altered by post-composition with an element of G so as to extend in sequence. Since given any deck transformation γ, $\mathbf{dev} \circ \gamma$ is again a developing map, it equals $h(\gamma) \circ \mathbf{dev}$ for $h(\gamma) \in G$. One can see easily that h is a homomorphism $\pi_1(M) \to G$. h is said to be the *holonomy homomorphism* and its image the *holonomy group*. Given an (X, G)-structure on M, $(\mathbf{dev}, h)$ is determined up to the equivalence relation $(\mathbf{dev}, h(\cdot)) \sim (\vartheta \circ \mathbf{dev}, \vartheta \circ h(\cdot) \circ \vartheta^{-1})$ where ϑ is an element of G. (Fundamental classes of examples are given by Sullivan-Thurston [**44**].)

A real projective space $\mathbf{R}P^n$ is given as the quotient space of $\mathbf{R}^{n+1} - \{O\}$ for the origin O by relation generated by scalar multiplications. The group of general linear transformations $\mathrm{GL}(n+1, \mathbf{R})$ descends to the group $\mathrm{PGL}(n+1, \mathbf{R})$ acting on $\mathbf{R}P^n$ as projective transformations. An affine space $\mathbf{R}^n$ is a Euclidean space $\mathbf{R}^n$ with the group of affine transformations $\mathrm{Aff}(\mathbf{R}^n)$ acting on it.

A *real projective n-manifold* is an n-manifold with an $(\mathbf{R}P^n, \mathrm{PGL}(n+1, \mathbf{R}))$-structure; an *affine n-manifold* is one with an $(\mathbf{R}^n, \mathrm{Aff}(\mathbf{R}^n))$-structure.

A prominent feature of real projective or affine manifolds is that geodesics are defined by connections. A curve in a real projective manifold is *geodesic* if it corresponds to a straight line in the projective space under charts, and similarly for curves in affine manifolds. *Segments* or *lines* are imbedded images of geodesics.

For affine manifolds, the completeness of geodesics makes sense since an affine parameter of a geodesic is defined. By a *complete* real line, we mean a geodesic in an affine manifold which has affine parameter from $-\infty$ to ∞. An affine manifold is often incomplete even if it is closed (see Kobayashi [**33**], Kobayashi-Nagano [**34**], and Sullivan-Thurston [**44**].)

Let $\mathbf{R}P^{n-1}$ be an $(n-1)$-dimensional subspace of $\mathbf{R}P^n$. Then $\mathbf{R}P^n - \mathbf{R}P^{n-1}$ has a natural structure of an affine space $\mathbf{R}^n$ so that the subgroup of projective transformations acting on the space equals $\mathrm{Aff}(\mathbf{R}^n)$ and projective and affine geodesics agree on $\mathbf{R}P^n - \mathbf{R}P^{n-1}$ up to parameterization. Obviously, an affine structure on a manifold induces a unique real projective structure since charts to $\mathbf{R}^n$ are naturally charts to $\mathbf{R}P^n$ and affine transition maps are real projective ones by the above identification. We will always regard an affine manifold as a real projective manifold obtained in this natural manner.

We assume in this paper that the manifold-boundary δM of a real projective manifold M is empty or totally geodesic; i.e., for each point of δM, there exists an open neighborhood U and a chart $\phi : U \to \mathbf{R}P^n$ so that $\phi(U)$ does not meet the subspace $\mathbf{R}P^{n-1}$ and is a convex open domain intersected with a closed affine half-space in $\mathbf{R}P^n - \mathbf{R}P^{n-1}$. We also assume that affine manifolds have empty or totally geodesic boundary; i.e., they have empty or totally geodesic boundary as real projective manifolds.

We give some simple examples and a short history on real projective and affine manifolds relevant to this paper.

EXAMPLE 1.1. Let ϑ be a matrix with the absolute value of all its eigenvalues greater than one. The quotient of $\mathbf{R}^n - \{O\}$ by $\langle\vartheta\rangle$ is homeomorphic to $\mathbf{S}^{n-1} \times \mathbf{S}^1$, and has a natural radiant affine structure. The manifold is said to be a *Hopf manifold.*

If ϑ preserves an $(n-1)$-dimensional subspace, then the quotient of $H - \{O\}$ where H is a closed affine half-space bounded by the subspace is homeomorphic to an $(n-1)$-ball times $\mathbf{S}^1$. This is said to be a *half-Hopf* manifold. This has totally geodesic boundary.

Let $x_1, \dots, x_n$ be coordinates of $\mathbf{R}^n$. Let H be an open upper half-space with the positive x_n-axis removed. Let ϑ be given by

$$(x_1, \dots, x_{n-1}, x_n) \mapsto (2x_1, \dots, 2x_{n-1}, x_n)$$

and φ by

$$(x_1, \dots, x_{n-1}, x_n) \mapsto (2x_1, \dots, 2x_{n-1}, 2x_n).$$

Then $H/\langle\vartheta, \varphi\rangle$ is homeomorphic to the $\mathbf{S}^{n-2} \times \mathbf{S}^1 \times \mathbf{S}^1$, and has a radiant affine structure. Denote this radiant affine manifold by $\mathcal{E}_2$.

One of the simplest example of an annulus with a real projective structure is given as the quotient of an invariant triangle by an infinite cyclic group of real projective transformations with representing matrices conjugate to diagonal matrices with positive distinct eigenvalues; a compact annulus with interior real projectively homeomorphic to such an annulus is said to be an *elementary* annulus (see [**12**] and [**17**]).

The Euler characteristic zero closed surfaces with real projective structures were classified by Goldman [**23**] and Nagano-Yagi [**38**]; such a surface either admits a natural compatible affine structure or is built up from elementary annuli by gluing (see [**23**] and [**17**]). If a closed surface Σ with a real projective structure has a negative Euler characteristic, then it is shown that Σ is obtained from surfaces with convex real projective structures and elementary annuli by projective gluing (see [**17**], [**11**], [**12**], and [**13**]). Since Goldman [**23**] classified and constructed all convex real projective structures on surfaces up to isotopy, we can classify and construct all real projective structures on closed surfaces. The deformation space of real projective structures on Σ, i.e., the space of equivalence classes of real projective structures on Σ up to isotopy with appropriate topology, is shown to be homeomorphic to a countably infinite disjoint union of cells of dimension $16g - 16$ where g is the genus of Σ [**17**].

A source of examples of real projective manifolds is obtained from the Klein model of hyperbolic space: Given a Lorentzian metric on $\mathbf{R}^{n+1}$, the positive part of the hyperboloid $-x_0^2 + x_1^2 + \cdots + x_n^2 = -1$ is the hyperbolic n-space H^n, and the group $\mathrm{PSO}(1,n)$ acts on H^n as a group of isometries. Under the projection $\mathbf{R}^{n+1} - \{O\} \to \mathbf{R}P^n$, H^n becomes identified with an open ball B^n in $\mathbf{R}P^n$ and $\mathrm{PSO}(1,n)$ with the natural copy of $\mathrm{PSO}(1,n)$ in $\mathrm{PGL}(n+1,\mathbf{R})$. Therefore, it follows that a hyperbolic structure on a manifold M induces a real projective structure on M. When $n = 2$, Koszul [**36**], Kac-Vinberg [**28**], and Goldman [**23**] found parameters of nontrivial deformations of hyperbolic real projective structures to non-hyperbolic but convex ones. It is thought that many such deformations exist in dimensions ≥ 3 giving us a plethora of real projective structures. Koszul found some higher dimensional deformations, and Goldman [**22**] parameters of deformations of real projective structures on the complement of the figure eight knot.

Many examples of affine manifolds were constructed by Sullivan-Thurston [**44**] from real projective manifolds and complex projective manifolds by suspension-like constructions. If a closed surface admits an affine structure, then its Euler characteristic is zero as shown by Benzécri [**7**]. The outstanding conjecture on affine n-manifolds is Chern's conjecture that the Euler characteristic of a closed affine manifold is zero. Kostant-Sullivan [**35**] showed this to be true if the affine manifold is complete, i.e., it is a quotient of $\mathbf{R}^n$ by a properly discontinuous and free action of a subgroup of $\mathrm{Aff}(\mathbf{R}^n)$. If the fundamental group of a closed affine manifold is amenable (or isomorphic to a free product of groups of polynomial growth), then Chern's conjecture is true by Hirsch-Thurston [**27**] (see also Kim-Lee [**31**], [**30**]). However, this conjecture is yet to be proved for general case, and strangely there seem to be really no tools to study this question.

Other examples of affine and projective manifolds were constructed by Smillie [**41**], in particular with diagonal holonomy, and later more general class were found by Benoist [**5**] and [**6**].

Another conjecture due to L. Markus states that an affine manifold is complete if and only if it has a volume form parallel with respect to the affine connection. This conjecture was settled if the fundamental group is virtually abelian, nilpotent, or solvable of rank $\leq n$ by Smillie [**42**] and Fried, Goldman, and Hirsch [**20**]. The conjecture is verified if the holonomy group lies in the group of Lorentzian transformations of $\mathbf{R}^n$ with a flat Lorentz metric by Carrière [**9**].

One question is to find an example of a three-manifold not admitting a real projective or affine structure, as asked by Goldman. Smillie [**43**] showed that a connected sum of lens spaces does not admit an affine structure. Choi [**16**] showed that an affine 3-manifold with parallel volume form must be irreducible and has a 3-cell as a universal cover, answering Carrière's question [**9**]. (See [**14**] for a summary.)

Going to the main part of this section, we now define the splitting of a real projective manifold M along a submanifold N of codimension one in M^o. We take a regular neighborhood K_i of each component S_i of N. If S_i is two-sided, then we take the closure in K_i of each component of $K_i - S_i$. If S_i is one-sided, then we take a double cover (K_i', p_i) of K_i corresponding to the index-two-subgroup of $\pi_1(K_i)$ given by the homomorphism $\pi_1(\delta K_i) \to \pi_1(K_i)$ induced by the inclusion map and remove S_i' the part corresponding to S_i in K_i. Take a component J of $K_i' - S_i'$. Then the closure J' of J has a natural real projective or affine structure with totally geodesic boundary equivalent to S_i'. J is obviously identical with $K_i - S_i$. For each component of $K_i - S_i$, we identify it with the appropriate subset of the closures as above to obtain the split manifold. Our construction gives us a real projective or affine manifold with totally geodesic boundary; the resulting manifold is said to be obtained from *splitting along* N.

There exists a standard double cover $p : \mathbf{S}^n \to \mathbf{R}P^n$ from the standard unit sphere $\mathbf{S}^n$ in $\mathbf{R}^{n+1}$ with a standard Riemannian metric μ which is projectively flat and has the same geodesic structure as one induced by p from $\mathbf{R}P^n$, i.e., μ-geodesics in $\mathbf{S}^n$ agree with projective geodesics in $\mathbf{S}^n$ up to parameterization. The metric μ also induces one on $\mathbf{R}P^n$, and the geodesic structure of this Riemannian metric agree with that of the given real projective structure on $\mathbf{R}P^n$ up to parameterization. Denote by $\mathbf{d}$ the distance metric on $\mathbf{S}^n$ induced from μ. $\mathbf{S}^n$ can also be identified with the quotient of $\mathbf{R}^{n+1} - \{O\}$ by the equivalence relation generated by positive scalar multiplications. The general linear group $\mathrm{GL}(n+1, \mathbf{R})$ acts on $\mathbf{R}^{n+1} - \{O\}$ and hence on its quotient space $\mathbf{S}^n$. The induced self-diffeomorphisms form a group $\mathrm{Aut}(\mathbf{S}^n)$ of all projective self-diffeomorphisms of $\mathbf{S}^n$, which obviously is isomorphic to the subgroup $\mathrm{SL}_{\pm}(n+1, \mathbf{R})$ of $\mathrm{GL}(n+1, \mathbf{R})$ consisting of linear maps with determinant ± 1.

A homeomorphism $f : X \to Y$ for metric spaces $(X, \mathbf{d}_X)$ and $(Y, \mathbf{d}_Y)$ are said to be *quasi-isometric* if

$$C^{-1}\mathbf{d}_X(x,y) \leq \mathbf{d}_Y(f(x), f(y)) \leq C\mathbf{d}_X(x,y), x, y \in X$$

hold for a uniform positive constant C independent of x and y. We introduce a Kuiper completion of $\tilde{M}$ (see [**37**]). Given a development pair $(\mathbf{dev}, h)$ of a real projective manifold M, $\mathbf{dev}$ lifts to an immersion $\mathbf{dev}' : \tilde{M} \to \mathbf{S}^n$ and h to $h' : \pi_1(M) \to \mathrm{Aut}(\mathbf{S}^n)$ so that they satisfy $h'(\gamma) \circ \mathbf{dev}' = \mathbf{dev}' \circ \gamma$ for each $\gamma \in \pi_1(M)$. We induce the Riemannian metric μ on $\tilde{M}$ from μ on $\mathbf{S}^n$, complete the distance metric $\mathbf{d}$ on $\tilde{M}$ induced from μ, and denote by $\check{M}$ the Cauchy completion of $\tilde{M}$ and $\mathbf{d}$ the completed metric. By an abuse of notation, we will simply denote $(\mathbf{dev}', h')$ by $(\mathbf{dev}, h)$. Since $\mathbf{dev}$ is distance non-increasing, $\mathbf{dev}$ extends to the completion $\check{M}$ and since each deck transformation ϑ is a quasi-isometry with respect to $\mathbf{d}$, it extends to a self-homeomorphism of $\check{M}$. We will denote the extended maps by same symbols $\mathbf{dev}$ and ϑ respectively (see [**29**] and [**15**]).

Given a real projective structure on M, any other development pair $(\mathbf{dev}'' : \tilde{M} \to \mathbf{S}^n, h''(\cdot))$ equals $(\vartheta \circ \mathbf{dev}, \vartheta \circ h(\cdot) \circ \vartheta^{-1})$ for an element $\vartheta \in \mathrm{Aut}(\mathbf{S}^n)$. The completion with respect to $\mathbf{dev}''$ is quasi-isometric to $\check{M}$ with $\mathbf{d}$ above since the metric induced from $\mathbf{dev}''$ is quasi-isometric with respect to the original μ. We will say that $(\check{M}, \mathbf{d})$ is a *Kuiper completion* of $\tilde{M}$.

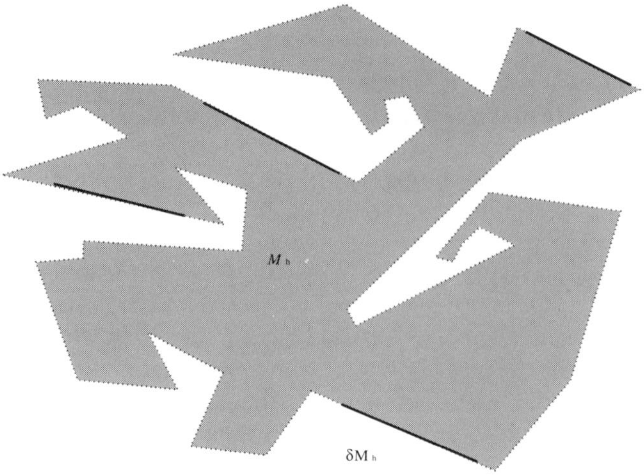

FIGURE 1.1. A figure of $\tilde{M}$. The dark lines indicate $\delta\tilde{M}$ and the dotted lines the ideal boundary $\tilde{M}_\infty$.

Recall that the complement in $\mathbf{R}P^n$ of a codimension-one subspace $\mathbf{R}P^{n-1}_\infty$ can be identified with $\mathbf{R}^n$, and the group of real projective transformations of $\mathbf{R}P^n$ acting on $\mathbf{R}P^{n-1}$ to $\mathrm{Aff}(\mathbf{R}^n)$ (see Berger [**8**]). The subspace $\mathbf{R}P^{n-1}_\infty$ corresponds to the great sphere $\mathbf{S}^{n-1}_\infty$ under the double covering map $\mathbf{S}^n \to \mathbf{R}P^n$, and $\mathbf{R}^n$ to an open hemisphere bounded by $\mathbf{S}^{n-1}_\infty$. Let $\mathcal{H}$ denote the closed hemisphere including this, and $\mathrm{Aut}(\mathcal{H})$ the group of orientation-preserving projective transformations acting on $\mathcal{H}$. Under the double covering map the interior $\mathcal{H}^o$ of $\mathcal{H}$ corresponds to the affine space $\mathbf{R}^n$ and $\mathrm{Aut}(\mathcal{H})$ to $\mathrm{Aff}(\mathbf{R}^n)$ in a one-to-one manner, preserving the geodesic structures up to parameterization. (We will fix our choice of the hemisphere $\mathcal{H}$ and the identification of $\mathcal{H}^o$ with $\mathbf{R}^n$, which determines of $\mathbf{S}^{n-1}_\infty$, and we use $\mathbf{R}^n$ and $\mathcal{H}^o$ interchangeably, and so use $\mathrm{Aff}(\mathbf{R}^n)$ and $\mathrm{Aut}(\mathcal{H})$.)

The correspondence between $\mathbf{R}^n$ and the open n-hemisphere is realized by moving the subspace $\mathbf{R}^n$ in $\mathbf{R}^{n+1}$ in the orthogonal direction by a unit and stereographically project from the origin onto a hemisphere $\mathbf{S}^n$ (see [**15**]).

Let M be an affine n-manifold. By above consideration, when $\mathcal{H}^o$ is identified with $\mathbf{R}^n$ and the affine transition functions with the real projective ones, M has a naturally induced real projective structure. Thus, there exists a development pair $(\mathbf{dev}, h)$ for M considered as a real projective manifold so that $\mathbf{dev} : \tilde{M} \to \mathcal{H}$ and $h(\gamma) \in \mathrm{Aut}(\mathcal{H})$ for $\gamma \in \pi_1(M)$, and clearly $\mathbf{dev}$ maps $\tilde{M}$ into $\mathcal{H}$.

We introduce the holonomy cover M_h of a real projective n-manifold M: Since the holonomy homomorphism may change only by a conjugation by an element of $\mathrm{Aut}(\mathbf{S}^n)$, the kernel K_M of the holonomy homomorphism h is independent of the choice of h. Considering $\pi_1(M)$ as the group of deck transformations, we define the *holonomy cover* M_h as the cover of M corresponding to K_M; that is, we define $M_h = \tilde{M}/K_M$ and identify K_M with $\pi_1(M_h)$. While we have $\mathbf{dev} \circ \vartheta = h(\vartheta) \circ \mathbf{dev} = \mathbf{dev}$ for any deck transformation ϑ in the kernel, it follows that $\mathbf{dev}$ induces a well-defined local diffeomorphism $\mathbf{dev}_h : M_h \to \mathbf{S}^n$. Given $\mathbf{dev}_h$, we can define a holonomy homomorphism $H_h : \pi_1(M)/\pi_1(M_h) \to \mathrm{Aut}(\mathbf{S}^n)$ from the deck-transformation group $\pi_1(M)/\pi_1(M_h)$ of M_h as we did for $(\mathbf{dev}, h)$; the resulting pair $(\mathbf{dev}_h, H_h)$ is said to be a *development pair*. Clearly H_h is induced homomorphism from h. Similarly to above, other development pair $(\mathbf{dev}'_h, H'_h(\cdot))$ equals $(\vartheta \circ \mathbf{dev}_h, \vartheta \circ H_h(\cdot) \circ \vartheta^{-1})$ for an element $\vartheta \in \mathrm{Aut}(\mathbf{S}^n)$. (For convenience, we drop the subscripts h from the notation $\mathbf{dev}_h$ from now on and write h for H_h.)

The immersion **dev** induces a Riemannian metric on M_h from $\mathbf{S}^n$, which we denote by μ, and μ induces a distance metric $\mathbf{d}$ on M_h; a completion of M_h is denoted by $\check{M}_h$ and the completed metric by $\mathbf{d}$; $M_{h\infty}$ denotes the ideal set $\check{M}_h - M_h$. As before, the developing map $\mathbf{dev} : M_h \to \mathbf{S}^n$ extends to a map $\check{M}_h \to \mathbf{S}^n$ and each deck transformation ϑ extends to a self-diffeomorphism $\check{M}_h \to \check{M}_h$. We will denote the extensions by **dev** and ϑ respectively.

EXAMPLE 1.2. The universal cover of a Hopf manifold M may be identified with $\mathbf{R}^n - \{O\}$, and $\check{M}_h$ with the n-hemisphere that is the closure of $\mathbf{R}^n - \{O\}$. The ideal set $M_{h\infty}$ is the union of $\{O\}$ and the sphere $\mathbf{S}^{n-1}$ which is the boundary.

The holonomy cover of $\mathcal{E}_2$ may be identified with $U - l$ where U is given by $x_n > 0$ and l the x_n-axis. $\check{\mathcal{E}}_{2,h}$ equals the closure of U in $\mathbf{S}^n$, and the ideal set the union of the boundary of U in $\mathbf{S}^n$ and l intersected with the closure of U.

A segment or a line in $\mathbf{S}^n$ is said to be *convex* if its $\mathbf{d}$-length is $\leq \pi$. A segment of $\mathbf{d}$-length $\leq \pi$ is always mapped to one of $\mathbf{d}$-length $\leq \pi$ by projective automorphisms of $\mathbf{S}^n$. We define a *convex subset* of $\mathbf{S}^n$ as a subset such that any two points of the set can be connected by a segment in the set of $\mathbf{d}$-length $\leq \pi$.

A *great* 0*-sphere* is the set of antipodal points, which is not convex. A *simply convex* subset of $\mathbf{S}^n$ is a convex subset of a $\mathbf{d}$-ball of radius $< \pi/2$, i.e., it is a precompact subset of an open hemisphere, and when the open hemisphere is identified with an affine space, it is a bounded affinely convex set. (See Figures 1.2 and 1.3).

It is easy to categorize all compact convex subsets of $\mathbf{S}^n$; they are homeomorphic to an i-ball or an i-sphere always. A compact convex subset of $\mathbf{S}^n$ is either a *great i-sphere* $i \geq 1$, i.e., a totally geodesic i-sphere; an *i-hemisphere*, i.e., the closure of a component of a great i-sphere with a great $(i-1)$-sphere in it removed; or a convex, proper compact subset of an i-hemisphere. Hence a compact convex subset is always homeomorphic to an i-ball or an i-sphere, and hence is a manifold, and we can define the boundary δA of a compact convex subset A. A convex subset of $\mathbf{S}^n$ is either a great i-sphere or a convex subset of a convex i-ball for some integer i. A convex, proper compact subset of an i-hemisphere contains a unique great sphere of dimension j, $0 \leq j < i$ or is a compact simply convex i-ball. (See [**15**] for details.)

See Chapter 6 of Ratcliffe [**39**] for more details on convex sets. Note that our definition of convexity is slightly different from the book. Assuming that a subset A of $\mathbf{S}^3$ is not a pair of antipodal points, a subset A is convex if and only if it is "convex" in the sense of Ratcliffe [**39**]. A pair of antipodal points is "convex" according to the book. A minor difficulty is that the intersection of two convex sets may not be convex according to our definition. But most of the theory passes to ours by mild and obvious modifications, which we will not endeavor to list.

A *side* of a convex compact subset C of $\mathbf{S}^n$ is a nonempty, maximal convex subset of δC. A *convex polyhedron* P in $\mathbf{S}^n$ is a nonempty, compact, convex subset of $\mathbf{S}^n$ such that the collection of its sides is finite. A side of a convex polyhedron of dimension n is again a convex polyhedron of dimension $n-1$. A convex polyhedron is always homeomorphic to a ball or equal to a great i-sphere, $i \geq 1$.

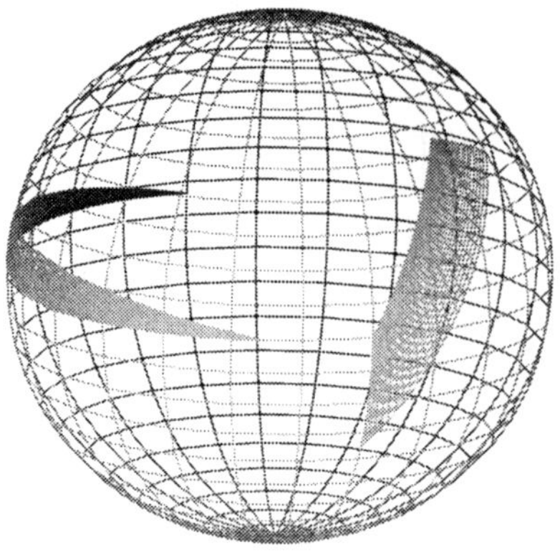

FIGURE 1.2. Some examples of convex sets in $\mathbf{S}^2$. The left one indicates a lune and the right one a simply convex set.

When our dimension n is 3, we will use the terms as follows: A *convex polygon* in $\mathbf{S}^3$ is a convex disk in a great 2-sphere with finitely many sides, and a *convex polyhedron* in $\mathbf{S}^3$ is a convex 3-ball in $\mathbf{S}^3$ with finitely many sides. A *lune* is the closure of a component of a great 2-sphere in $\mathbf{S}^3$ (or sometimes in $\mathbf{S}^2$) with two distinct great circles removed. A *triangle* is the closure of a component of one with three distinct great circles in general position removed.

A convex but not simply convex polyhedron in $\mathbf{S}^3$ is either a 3-hemisphere; a *bihedron*, i.e., the closure of a component of $\mathbf{S}^3$ with two distinct great 2-spheres removed; or a *tube over a k-gon* for $k \geq 3$, i.e., the closure of an appropriate component of $\mathbf{S}^3$ with k distinct great 2-spheres all containing a common pair of antipodal points removed. Any other convex polyhedron in $\mathbf{S}^3$ is a simply convex polyhedron. Since they form a bounded subset of the open hemisphere identified with an affine 3-space, the usual theory of Euclidean convex polyhedra applies.

The only convex polyhedron in $\mathbf{S}^3$ with three sides is a tube over a triangle for which we reserve the term *trihedron*; for a simply convex polyhedron with four side, we reserve the term *tetrahedron*.

DEFINITION 1.1. Given a subsets K of $\mathbf{S}^{n-1}_\infty$ and the origin O in $\mathcal{H}$, we say that the union of all segments starting from O and ending at points of K is the *cone over K*.

REMARK 1.1. It is easy to see that a cone over a compact convex subset K of $\mathbf{S}^{n-1}_\infty$ is a compact convex set. We also have the correspondence that K is a simply convex i-ball if and only if the cone over K is a simply convex $(i+1)$-ball. Moreover K is a polyhedron of dimension i if and only if the cone over K is a polyhedron of dimension $i+1$.

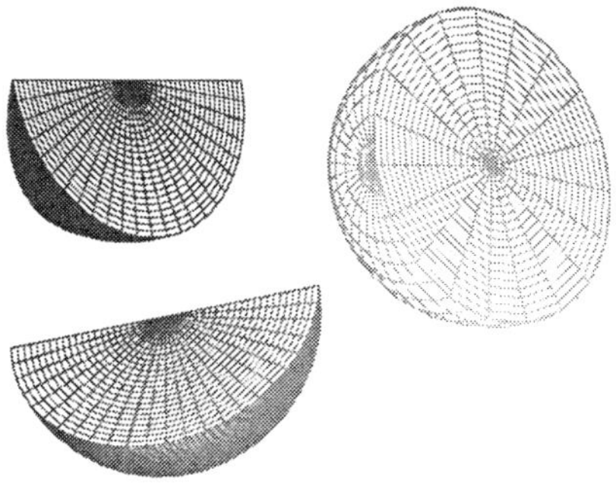

FIGURE 1.3. Some examples of convex sets in $\mathbf{S}^3$ stereographically projected to $\mathbf{R}^3$ and translated: Two trihedra and one bihedron. Some of their sides are drawn as totally geodesic sides for computer graphics convenience.

A *convex segment* in $\check{M}_h$ is a subset A such that $\mathbf{dev}|A$ is an imbedding onto a convex segment in $\mathbf{S}^n$. Given a subset A of $\check{M}_h$, we say that A is convex if two points of A can be connected by a convex segment in A. A *tame* subset of $\check{M}_h$ is a convex subset of M_h or a convex subset of a compact convex subset of $\check{M}_h$. The closure in $\check{M}_h$ of a convex subset of M_h is always tame since for any convex subset A of M_h, $\mathbf{dev}|A$ is a $\mathbf{d}$-isometry onto $\mathbf{dev}(A)$, and, hence, $\mathbf{dev}$ restricted to the closure $\mathrm{Cl}(A)$ of A is a $\mathbf{d}$-isometry onto $\mathrm{Cl}(\mathbf{dev}(A))$. Since $\mathbf{dev}|A$ is a $\mathbf{d}$-isometry for a tame set A, we also see that $\mathbf{dev}|A$ for a tame set A is always an imbedding onto $\mathbf{dev}(A)$.

Note that M is *convex* if M_h is a convex subset of $\check{M}_h$. Unfortunately, this definition is not equivalent to one in the introduction; however, it is almost the same (see Theorem A.2 in [**15**]) as convex subsets of $\mathbf{S}^n$ are almost always in an open hemisphere. (Recall that an open hemisphere corresponds to an affine patch.) For affine manifolds, the definitions are equivalent. In this paper, there is no confusion since we study only real projective manifolds coming from affine manifolds.

DEFINITION 1.2. We define an *i-ball* A in $\check{M}_h$ to be a compact subset of $\check{M}_h$ such that $\mathbf{dev}|A$ is a homeomorphism to an i-ball (not necessarily convex) in a great i-sphere and its manifold interior A^o is a subset of M_h. A *great i-sphere* A in $\check{M}_h$ is a compact subset of $\check{M}_h$ such that $\mathbf{dev}|A$ is a homomorphism onto a great i-sphere in $\mathbf{S}^n$.

As in [**15**], a topological i-ball is defined as above but A^o is not required to be in M_h.

We define *convex segments*, *convex sets*, *tame sets*, an *i-balls* in $\check{M}$ in the exactly the same manner as in $\check{M}_h$.

A convex *polyhedron* in $\check{M}_h$ is a topological i-ball A such that $\mathbf{dev}(A)$ is a polyhedron. A convex *polyhedron* in $\check{M}$ is defined similarly.

A compact convex subset A of $\check{M}_h$ is tame, and hence is either a topological i-ball or a great i-sphere for some i.

When the dimension of M equals 3, we will use terms in the following manner. A *convex polyhedron* A in $\check{M}_h$ is a convex 3-ball in $\check{M}_h$ whose image is a convex polyhedron in $\mathbf{S}^3$. A *convex polygon* in $\check{M}_h$ is a convex topological 2-ball of M_h whose image is a convex polygon in a great sphere in $\mathbf{S}^3$. These include lunes and triangles. A *side* of the polyhedron or polygon is defined obviously as above. (Note that hemispheres, bihedra, trihedra, tetrahedra in $\check{M}_h$ are polyhedra defined in this sense.) These definitions also apply to subsets of $\check{M}$ in the obvious manner.

REMARK 1.2. Note that all definitions here are well-defined under the change of development pairs and our definition of convexity is directly influenced by the fact that we require **dev** restricted to relevant convex sets to be imbeddings to $\mathbf{S}^n$.

The following theorem and lemma are true both in $\check{M}$ and $\check{M}_h$ as there are no difference in proofs. Let F_1 and F_2 be two convex n-balls in $\check{M}_h$. We say that F_1 and F_2 *overlap* if $F_1^o \cap F_2 \neq \emptyset$, which is equivalent to $F_1 \cap F_2^o \neq \emptyset$ or $F_1^o \cap F_2^o \neq \emptyset$.

PROPOSITION 1.1. *If F_1 and F_2 overlap, then* $\mathbf{dev}|F_1 \cup F_2$ *is an imbedding onto* $\mathbf{dev}(F_1) \cup \mathbf{dev}(F_2)$ *and* $\mathbf{dev}|F_1 \cap F_2$ *onto* $\mathbf{dev}(F_1) \cap \mathbf{dev}(F_2)$*. Moreover, $F_1 \cup F_2$ is an n-ball, and $F_1 \cap F_2$ is a convex n-ball.*

PROOF. See Proposition 3.9 in [**15**]. □

REMARK 1.3. If F_1 and F_2 be convex open balls in M_h and M is affine, the above theorem was proved by Carrière in Proposition 1.3.1 of [**9**]. The above theorem implies this result since any convex open n-ball in M_h has the closure which is a compact n-ball.

The above proposition follows from:

PROPOSITION 1.2. *Let A be a k-ball in $\check{M}_h$, and B an l-ball. Suppose that $A^o \cap B^o \neq \emptyset$,* $\mathbf{dev}(A) \cap \mathbf{dev}(B)$ *is a compact manifold in $\mathbf{S}^n$ with interior equal to* $\mathbf{dev}(A)^o \cap \mathbf{dev}(B)^o$*, and* $\mathbf{dev}(A)^o \cap \mathbf{dev}(B)^o$ *is arcwise-connected. Then* $\mathbf{dev}|A \cup B$ *is a homeomorphism onto* $\mathbf{dev}(A) \cup \mathbf{dev}(B)$.

PROOF. See Proposition 3.10 in [**15**]. □

CHAPTER 2

$(n-1)$-convexity: previous results

In this chapter, we summarize the facts that we need from the monograph [**15**]: We will define m-convexity of real projective manifolds, and explain that the failure of $(n-1)$-convexity of a real projective manifold implies the existence of n-crescents in the completion $\check{M}_h$ of the holonomy cover M_h of M. Finally, we explain the decomposition of M into $(n-1)$-convex manifolds and concave affine n-manifolds along $(n-1)$-dimensional manifold convex from the $(n-1)$-convex sides. An intermediate step of splitting the real projective manifold along two-faced $(n-1)$-manifolds, if they exist, is explained. (Two-faced submanifolds are closed submanifolds which arise from collections of n-crescents meeting at their boundary exclusively.)

Let M be a real projective n-manifold and M_h its holonomy cover. An *m-simplex* T in $\check{M}_h$ is a tame subset of $\check{M}_h$ such that $\mathbf{dev}|T$ is an imbedding onto an affine m-simplex in an affine patch in $\mathbf{S}^n$.

DEFINITION 2.1. We say that M is *m-convex*, $0 < m < n$, if the following holds: If $T \subset \check{M}_h$ be an $(m+1)$-simplex with sides $F_1, F_2, \dots, F_{m+2}$ such that $T^o \cup F_2 \cup \dots \cup F_{m+2} \subset M_h$, then $T \subset M_h$.

LEMMA 2.1. *M is m-convex if and only if $\check{M}_h$ includes no $(m+1)$-simplex T with a side F_1 such that $T \cap M_{h\infty} = F_1^o \cap M_{h\infty} \neq \emptyset$.*

PROOF. See Proposition 4.3 in [**15**]. □

LEMMA 2.2. *M is m-convex if and only if for an $(m+1)$-simplex T with sides $F_1, \dots, F_{m+2}$, every real projective immersion $f : T^o \cup F_2 \cup \dots \cup F_{m+2} \to M$ extends to one from T.*

PROOF. See Proposition 4.2 in [**15**]. □

It is easy to see that i-convexity implies j-convexity whenever $i \leq j < n$ (see Remark 2 in [**16**]). A compact real projective manifold with totally geodesic boundary is convex if and only if it is 1-convex (see Theorem A.2 of [**15**]).

For example, Hopf manifolds and half-Hopf manifolds are not $(n-1)$-convex. $\mathcal{E}_2$ is $(n-1)$-convex but not $(n-2)$-convex for $n \geq 3$ (see Example 1.1).

A bihedron in $\check{M}_h$ is said to be an *n-crescent* if one of its sides is a subset of $M_{h\infty}$ and the other side is not. An n-hemisphere A in $\check{M}_h$ is said to be an *n-crescent* if an $(n-1)$-hemisphere in δA is a subset of $M_{h\infty}$ and the boundary itself is not a subset of $M_{h\infty}$. *Bihedral* n-crescents are the former ones, and *hemispheric* n-crescents are the latter ones.

Given a bihedral n-crescent R, the interior of the side in $M_{h\infty}$ is denote by α_R and the other side by ν_R. Given a hemispheric n-crescent S, the union of all open $(n-1)$-hemispheres in $M_{h\infty} \cap \delta S$ is denoted by α_S and its complement in δS by

ν_S. Note that α_S is homeomorphic to a connected open manifold and ν_S is a tame set homeomorphic to an $(n-1)$-ball. (Note that we define n-crescents and their objects in $\check{M}$ in the same manner.)

Assume that M_h is not projectively diffeomorphic to an open n-bihedron or an open n-hemisphere. We proved in [**15**] the following theorem:

THEOREM 2.1. *Let M be a compact radiant affine manifold with empty or totally geodesic boundary. Suppose that M is not $(n-1)$-convex. Then $\check{M}_h$ includes an n-crescent.*

A *nice* $(n-1)$-submanifold of M is a totally geodesic closed real projective $(n-1)$-submanifold in M^o each component of which is projectively diffeomorphic to a quotient of a connected open subset of an affine patch $\mathbf{R}^{n-1}$ in the real projective space $\mathbf{R}P^{n-1}$ by a properly discontinuous and free action of a group of projective transformations.

If two hemispheric n-crescents R and S overlap, then they are equal. If R and S meet but do not overlap, then $R \cap M_h$ and $S \cap M_h$ meet in common components of $\nu_R \cap M_h$ and $\nu_S \cap M_h$. Such components are called a *copied* components, and the union of all such components for all hemispheric n-crescents is a properly imbedded submanifold in M_h and hence covers a closed submanifold in M^o, which is said to be a *two-faced submanifold arising from hemispheric n-crescents.*

We say that $R \sim S$ for two bihedral n-crescents R and S if R and S overlap. This generates an equivalence relation which we denote by $\sim$ again. We defined $\Lambda(R)$ to be $\bigcup_{S \sim R} S$ and discussed the properties of $\Lambda(R)$ in [**15**]. (See Chapter 4 discussing this in a more special setting.) The so-called two-faced $(n-1)$-submanifold arising from bihedral n-crescents is a properly imbedded submanifold in M covered by the submanifold of M_h that is the union of common components of the topological boundaries $\mathrm{bd}\Lambda(R) \cap M_h$ and $\mathrm{bd}\Lambda(S) \cap M_h$ for all pair of non-equivalent n-crescents R and S. Of course, such a set may be empty and a two-faced submanifold doesn't exist for our real projective manifold M.

A two-faced submanifold is a nice submanifold. We will repeat the construction in [**15**] in Chapter 4 in the radiant affine case, as needed by this paper.

DEFINITION 2.2. A *concave affine n-manifold M of type I* is a real projective manifold such that M_h is a subset of a hemispheric n-crescent in $\check{M}_h$. A *concave affine n-manifold M of type II* is a real projective manifold with possibly concave boundary such that $\check{M}_h$ includes no hemispheric n-crescents and M_h is a subset of $\Lambda(R)$ for a bihedral n-crescent R in $\check{M}_h$.

THEOREM 2.2. *Suppose that M is compact and $\check{M}_h$ includes a hemispheric n-crescent. Then M includes a compact concave affine n-submanifold N of type I or M^o includes the two-faced $(n-1)$-submanifold arising from hemispheric n-crescents.*

THEOREM 2.3. *Suppose that M is compact and $\check{M}_h$ includes bihedral n-crescents but no hemispheric n-crescents. Then M includes a compact concave affine n-submanifold N of type II or M^o includes the two-faced $(n-1)$-submanifold arising from bihedral n-crescents.*

A real projective manifold has *convex boundary* if each of its boundary points has a neighborhood where the chart restricted to it is an imbedding onto a convex set. A real projective manifold has *concave boundary* if each of its boundary points has a neighborhood where the chart restricted to it is an imbedding onto the complement of an open convex set inside a simply convex open ball.

The manifold interior of a concave affine submanifold always admits a compatible affine structure (see **[15]**).

The following corollary was stated and proved in **[15]** in a slightly different form. The proof in Chapter 10 of **[15]** easily shows:

COROLLARY 2.1. *Suppose that M is compact but not $(n-1)$-convex. Then*

- *after decomposing M along the two-faced $(n-1)$-submanifold A_1 arising from hemispheric n-crescents, the resulting manifold N decomposes into concave affine manifolds of type I and real projective n-manifolds with totally geodesic boundary the Kuiper completion of whose holonomy cover includes no hemispheric n-crescents.*
- *We let N^1 be the disjoint union of the resulting manifolds of the decomposition other than concave affine ones of type I. After splitting N^1 along the two-faced $(n-1)$-manifold A_2 arising from bihedral n-crescents, the resulting manifold $N^{1\prime}$ decomposes into concave affine manifolds of type II and real projective n-manifolds with convex boundary which is $(n-1)$-convex. The Kuiper completions of the holonomy covers of $(n-1)$-convex pieces include no n-crescents.*

Note that A_1 and A_2 could be empty. If $A_1 = \emptyset$, then we define $N = M$ and if $A_2 = \emptyset$, then we define $N^{1\prime} = N^1$. The proof depends on the following proposition which we will be using later.

REMARK 2.1. The manifolds A_1 and A_2 are nice. So are the boundary components of a concave affine manifold of type I since the manifold-boundary points of the universal cover of a concave affine manifold of type I lie in ν_R for a bihedral n-crescent R.

PROPOSITION 2.1. *Let A be a submanifold and a closed subset of a regular cover $\hat{N}$ of a compact n-manifold N so that for each deck transformation ϑ, $\vartheta(A)$ either equals A or is disjoint from A. Suppose that the collection composed of elements $\vartheta(A)$ for $\vartheta \in \mathrm{Aut}_N(\hat{N})$ for the deck-transformation group $\mathrm{Aut}_N(\hat{N})$ is locally finite, i.e., each point of $\hat{N}$ has a neighborhood that meets only finitely many members of the collection. Then $p|A : A \to p(A)$ for the covering map $p : \hat{N} \to N$ is a covering map onto a compact manifold A. Furthermore, if A is of dimension $< n$, and is an open manifold, then $p(A)$ is a closed submanifold of same dimension. If A is of dimension n, and has boundary δA, then $p(A)$ is a compact submanifold of dimension n with boundary $p(\delta A)$.*

PROOF. The readers can easily supply the proof. □

CHAPTER 3

Radiant vector fields, generalized affine suspensions, and the radial completeness

We will discuss general facts about radiant affine n-manifolds in this chapter: Given a radiant affine manifold M, the radiant vector field in $\mathbf{R}^n - \{O\}$ induces a radiant vector field on M_h and M. As examples of radiant affine n-manifolds, we define generalized affine suspensions over real projective surfaces and orbifolds (see Carrière [**10**] and Barbot [**3**]). We show that a radiant affine manifold is a generalized affine suspension if and only if it admits a total cross-section to the radial flow. The results of Barbot [**3**] that we need are stated here.

From an estimation of the radial flow in terms of **d**, it follows that the radial flow extends to the Kuiper completion as well, and the point on $\check{M}_h$ corresponding to the origin under **dev** is unique. The ideal sets are also radial-flow-invariant, and the completion of the holonomy cover of a radiant affine n-manifold, if not $(n-1)$-convex, must contain a radiant n-crescent, i.e., a radiant n-dimensional bihedron. Finally, we say about an intersection property of two radiant n-bihedra.

Let $e_0, e_1, \ldots, e_n$ be the orthonormal basis of $\mathbf{R}^{n+1}$ and $x_0, x_1, \ldots, x_n$ the associated coordinates, and $\mathbf{S}^n$ the unit sphere in $\mathbf{R}^{n+1}$ with origin O. Let ϕ denote the radial projection from $\mathbf{R}^{n+1} - \{O\}$ to $\mathbf{S}^n$. For convenience, we identify $\mathbf{R}^n$ with the interior $\mathcal{H}^o$ of the upper hemisphere $\mathcal{H}$ in $\mathbf{S}^n$ by sending $x \mapsto x + e_0$ and then sending it to $\mathbf{S}^n$ by ϕ. This identification is compatible with standard affine and real projective structures on $\mathbf{R}^n$ and $\mathcal{H}^o$ as given in the introduction, i.e., geodesic structures are preserved. (Note that the origin O of $\mathbf{R}^n$ corresponds to e_0.)

We define the radiant vector field $\mathbf{V}$ on $\mathbf{R}^n$ to be given by $\sum_{i=1}^n x_i \partial/\partial x_i$, where $\mathbf{V}(O)$ is the zero vector. The vector field $\mathbf{V}$ on $\mathcal{H}$ generates a flow. It is easy to compute that $\mathbf{V}$ is a vector field on $\mathcal{H}$ of bounded **d**-length and $\mathbf{V}$ extends to a continuous vector field on $\mathbf{S}^n$ by a zero vector field on $\mathbf{S}^n - \mathcal{H}$.

Since $\mathbf{V}$ is a **d**-bounded vector field, the flow Φ'_t on $\mathbf{S}^n$ generated by $\mathbf{V}$ satisfies the following inequality:

$$C(t)^{-1}\mathbf{d}(x,y) \leq \mathbf{d}(\Phi'_t(x), \Phi'_t(y)) \leq C(t)\mathbf{d}(x,y) \tag{3.1}$$

for $x, y \in \mathcal{H}^o - O$ and a positive constant $C(t)$ smoothly depending only on t. (See Lemma 3 of Section 2.1 of Abraham-Marsden[**1**].)

Let M be a compact radiant affine n-manifold with a developing map $\mathbf{dev}: M_h \to \mathbf{R}^n$ and the holonomy homomorphism $h: \pi_1(M)/\pi_1(M_h) \to \mathrm{Aff}(\mathbf{R}^n)$ and the holonomy group Γ equal to $h(\pi_1(M))$, fixing the origin O. As $\mathbf{V}$ on $\mathcal{H}$ is Γ-invariant, the induced vector field $\mathbf{V}_h$ in M_h by **dev** is invariant under the deck transformations; hence, there exists an induced vector field $\mathbf{V}_M$ on M called the *radiant vector field* of M, which defines a flow $\Phi: \mathbf{R} \times M \to M$. Since M is a closed manifold, Φ is a complete flow. The vector field $\mathbf{V}_h$ induces the flow

$$\Phi_h : \mathbf{R} \times M_h \to M_h$$

so that the following diagram commutes for every $t \in \mathbf{R}$

$$\begin{array}{ccc} M_h & \xrightarrow{\Phi_{h,t}} & M_h \\ \mathbf{dev} \downarrow & & \downarrow \mathbf{dev} \\ \mathbf{R}^n - \{O\} & \xrightarrow{\Phi'_t} & \mathbf{R}^n - \{O\}. \end{array} \tag{3.2}$$

A *radial line* in $\mathbf{R}^n - \{O\}$ is a component of a complete line with O removed.

LEMMA 3.1. *The images* $\mathbf{dev}(M_h)$ *and* $\mathbf{dev}(\tilde{M})$ *miss* O, *and for each point* x *of* M_h, *there exists a unique imbedded geodesic* l, *maximal in* M_h, *passing through* x *such that* $\mathbf{dev}|l$ *is a diffeomorphism to a radial line in* $\mathbf{R}^n$.

PROOF. The first statement is proved by Theorem 3.3 of [**20**]. The last statement follows from the completeness of the flow Φ_h. □

In the above theorem, the line l is said to be the radial line through x.

We now define generalized affine suspensions over real projective manifolds and orbifolds (see Carrière [**10**] and Barbot [**3**]). A *real projective orbifold* is simply an orbifold with geometric structure modeled on $(\mathbf{R}P^n, \mathrm{PGL}(n+1, \mathbf{R}))$. (For definitions of orbifold and geometric structures on orbifolds, see Ratcliffe [**39**].)

Let Σ be a compact real projective $(n-1)$-manifold with empty or totally geodesic boundary with a projective automorphism ϕ. Since $\mathbf{S}^{n-1}_\infty$ is a real projective $(n-1)$-sphere, there exists a developing map $\mathbf{dev} : \tilde{\Sigma} \to \mathbf{S}^{n-1}_\infty$ and a holonomy homomorphism $h : \pi_1(\Sigma) \to \mathrm{Aut}(\mathbf{S}^{n-1}_\infty)$. Choosing an arbitrary Euclidean metric in $\mathbf{R}^n$, we define an immersion $\mathbf{dev}' : \tilde{\Sigma} \times \mathbf{R} \to \mathbf{R}^n$ by simply mapping (x, t) to $e^t u(x)$ where $u(x)$ is the unit vector at the origin in the direction of $\mathbf{dev}(x)$ in $\mathbf{R}^n$. Since $\mathrm{Aut}(\mathbf{S}^{n-1}_\infty)$ can be identified with $\mathrm{SL}_\pm(n, \mathbf{R})$, there is a natural quotient map $\mathrm{GL}(n, \mathbf{R}) \to \mathrm{Aut}(\mathbf{S}^{n-1}_\infty)$; we choose any lift $h' : \pi_1(\Sigma) \to \mathrm{GL}(n, \mathbf{R})$ of h, and define a corresponding action of $\pi_1(\Sigma)$ on $\tilde{\Sigma} \times \mathbf{R}$ by $\vartheta(x, t) = (\vartheta(x), t + \log ||h'(\vartheta)(u(x))||)$.

Note that there is a unique lift $h'' : \pi_1(\Sigma) \to \mathrm{SL}_\pm(n, \mathbf{R})$ and $h'(\varphi) = k(\varphi)h''(\phi)$ for a homomorphism $k : \pi_1(\Sigma) \to \mathbf{R}^+$. Clearly k induces a homomorphism $k_\# : H_1(\Sigma) \to \mathbf{R}$ by taking logarithms. We choose h' to satisfy

$$k_\# \circ \phi_\# = \phi_\# \tag{3.3}$$

for induced homomorphism $\phi_\# : H_1(\Sigma) \to H_1(\Sigma)$ of ϕ.

Letting $\tilde{\Sigma} \times \mathbf{R}$ have the affine structure induced from the immersion $\mathbf{dev}'$, we see that $\pi_1(\Sigma)$ defines a properly discontinuous and free affine action of $\tilde{\Sigma} \times \mathbf{R}$ preserving each fiber homeomorphic to $\mathbf{R}$, and the quotient space is homeomorphic to $\Sigma \times \mathbf{R}$, i.e., a trivial $\mathbf{R}$-fiber bundle over Σ. We identify the quotient space with $\Sigma \times \mathbf{R}$, and choose a section $s : \Sigma \to \Sigma \times \mathbf{R}$ so that $s(\Sigma)$ becomes a compact imbedded surface.

The projective automorphism ϕ lifts to a projective automorphism $\tilde{\phi}$ of $\tilde{\Sigma}$. Since $\tilde{\phi}$ is a projective automorphism, there exists an element ρ in $\mathrm{Aut}(\mathbf{S}^{n-1}_\infty)$ satisfying $\mathbf{dev} \circ \tilde{\phi} = \rho \circ \mathbf{dev}$. We may choose any element ρ' of $\mathrm{GL}(n, \mathbf{R})$ which induces ρ, and ρ' defines an affine automorphism ρ'' of $\tilde{\Sigma} \times \mathbf{R}$ given by $\rho''(x, t) = (\tilde{\phi}(x), t + \log ||\rho'(u(x))||)$.

Given a deck transformation ϑ of $\tilde{\Sigma}$, there exists a deck transformation φ satisfying $\phi \circ \vartheta = \varphi \circ \phi$. φ is to be denote by $\phi^*(\vartheta)$ and $\phi^* : \pi_1(\Sigma) \to \pi_1(\Sigma)$ defines a group automorphism. By equation 3.3, we see that $\rho'' \circ \vartheta = \varphi \circ \rho''$ on $\tilde{\Sigma} \times \mathbf{R}$. Therefore, it follows that ρ'' induces an affine automorphism $a_{\rho'}$ of $\Sigma \times \mathbf{R}$.

We let $e^r\mathrm{I}$ for $r \in \mathbf{R}$ denote the *dilatation* multiplying each vector in $\mathbf{R}^3 - \{O\}$ by a factor e^r, which induces an affine automorphism D_r on $\tilde{\Sigma} \times \mathbf{R}$ also called a *dilatation* given by $D_r(x,t) = (x, t+r)$. Since D_r commutes with any deck transformation of $\tilde{\Sigma} \times \mathbf{R}$, it follows that D_r defines an affine fiber-preserving automorphism D_r' of $\tilde{\Sigma} \times \mathbf{R}$.

Any other choice ρ_1' in $\mathrm{GL}(n, \mathbf{R})$ of ρ' equals $e^r\mathrm{I} \circ \rho'$ for some r, and given the affine automorphism $a_{\rho_1'}$ of $\Sigma \times \mathbf{R}$ corresponding to ρ_1', we see that $a_{\rho_1'}$ equals $D_r \circ a_{\rho'}$ for some r. Hence by choosing r sufficiently large > 1, and positive, we can make $a_{\rho_1'}(s(\Sigma))$ and $s(\Sigma)$ disjoint and $a_{\rho_1'}(s(\Sigma))$ to lie in the radially outer-component of $\Sigma \times \mathbf{R} - s(\Sigma)$. Since $a_{\rho_1'}$ is a fiber-preserving diffeomorphism, $a_{\rho_1'}(s(\Sigma))$ is another total cross-section. We let N denote the compact $(n+1)$-manifold in $\Sigma \times \mathbf{R}$ bounded by $s(\Sigma)$ and $a_{\rho_1'}(s(\Sigma))$, and identify $s(\Sigma)$ and $a_{\rho_1'}(s(\Sigma))$ by $a_{\rho_1'}$ to obtain a compact radiant affine n-manifold homeomorphic to the mapping torus $\Sigma \times_\phi \mathbf{S}^1$, i.e., $\Sigma \times I/\sim$ where $\sim$ is defined by $(x, 0) \sim (\phi(x), 1)$. We call the resulting affine n-manifold the *generalized affine suspension over* Σ *using* the projective automorphism ϕ. (Note the term "affine suspension" is reserved for the case Σ and ϕ are both affine.)

If we let ϕ be the identity automorphism of Σ, then the generalized affine suspension is a so-called Benzécri suspension. But even when ϕ is of finite order, we will also call it *Benzécri suspension* over the projective orbifold $\Sigma/\langle\phi\rangle$ in this paper (sometimes, over the manifold Σ also).

PROPOSITION 3.1. *A generalized affine suspension of a real projective $(n-1)$-manifold with totally geodesic or empty boundary is a radiant affine n-manifold with totally geodesic or empty boundary.*

PROOF. Straightforward. □

When $n = 3$, the boundary components are homeomorphic to tori or Klein bottles since they are tangent to the radial flow, and hence have zero Euler characteristic.

EXAMPLE 3.1. Let T be an $(n-1)$-dimensional Hopf manifold given as the quotient of $\mathbf{R}^{n-1} - \{O\}$ by the cyclic group generated by g sending $x \to 2x$ for each vector x. Clearly, $\mathcal{E}_2$ is a Benzécri suspension of T.

Given a closed projective surface Σ, we can obtain Benzécri suspensions easily. When Σ is a quotient of a standard ball in $\mathbf{R}P^2$; i.e., Σ has a projective structure induced from hyperbolic structure, then the Benzécri suspension can be obtained from the upper part of the interior of the null cone in the Lorentzian space $\mathbf{R}^{1,n-1}$ by an action of a group generated by linear Lorentz transformations and a homothety, i.e., a linear map of form $s\mathrm{I}$ for $s > 0$. In this case, our Benzécri suspension carries Lorentzian flat conformal structure as well.

Given an arbitrary closed real projective surface Σ of negative Euler characteristic, Σ decomposes along disjoint simple closed geodesics into convex real projective surfaces and annuli with geodesic boundary (see [**11**], [**12**], and [**13**]). Thus the Benzécri suspension also decomposes along corresponding totally geodesic tori into convex radiant affine manifolds and Benzécri suspensions of the annuli.

THEOREM 3.1. *Let Σ be a compact real projective surface with totally geodesic boundary. If the Euler characteristic of Σ is negative, then a generalized affine suspension of Σ using an automorphism ϕ is homeomorphic to a Seifert space with zero Euler number with base orbifold homeomorphic to $\Sigma/\langle\phi\rangle$ where ϕ is a finite order automorphism. Moreover, the generalized affine suspension has a finite cover which is a Benzécri suspension over Σ.*

PROOF. Since by following Theorem 3.2, ϕ is of finite order, it follows from the fact that M is a mapping torus of ρ over Σ that M is homeomorphic to a Seifert manifold over $\Sigma/\langle\phi\rangle$.

Since a finite power of ϕ is the identity automorphism of Σ, it follows that M is finitely covered by a Benzécri suspension, which is homeomorphic to a trivial circle bundle over Σ. Since the Euler number of the trivial circle bundle is zero, the conclusion follows. □

A projective automorphism of $\mathbf{S}^2$ is said to be *hyperbolic* if it is represented by a diagonal matrix with distinct positive eigenvalues in $\mathrm{GL}(3, \mathbf{R})$. It is said to be *quasi-hyperbolic* if it is represented by a nondiagonal matrix with two distinct positive eigenvalues (see [**12**] for details). Let S be a compact real projective surface with empty or geodesic boundary and negative Euler characteristic. Let $(\mathbf{dev}, h)$ be its development pair. A *strong tight curve* in a real projective surface S is a simple closed geodesic α such that its lift $\tilde{\alpha}$ in $\tilde{S}$ is a simply convex line and $h(\vartheta)$ for the deck transformation ϑ corresponding to $\tilde{\alpha}$ and α is hyperbolic or quasi-hyperbolic, and $\mathbf{dev}(\tilde{\alpha})$ is a geodesic connecting the fixed point of the largest eigenvalue to that of the smallest eigenvalue of $h(\vartheta)$.

REMARK 3.1. If the Euler characteristic of S is zero, then the generalized affine suspension of Σ is not necessarily Seifert. Actually there are many nontrivial automorphisms which are not finite order up to isotopy.

THEOREM 3.2. *Let $\rho : S \to S$ be a projective automorphism of a compact real projective surface S of negative Euler characteristic. Then ρ is of finite order.*

PROOF. A *purely convex* real projective surface L is a compact convex surface with totally geodesic or empty boundary and negative Euler characteristic such that it includes no compact annulus with geodesic boundary freely homotopic to a boundary component of L. The real projective surface S canonically decomposes into maximal purely convex real projective subsurfaces of negative Euler characteristics and maximal annuli or Möbius bands with geodesic boundary [**12**], [**13**]. By uniqueness of the decomposition in [**13**], ρ acts on the union of the purely convex surfaces which have negative Euler characteristics and are mutually disjoint. Letting S' denote the disjoint union, we see that each boundary component of S' is a strong tight curve by Proposition 4.5 in [**12**].

Assume that ρ is of infinite order. By taking a power of ρ if necessary, we assume that ρ acts on every component and boundary component of S'. We assume without loss of generality that S' is connected. Since ρ acts on a boundary component k of S', ρ lifts to an automorphism $\tilde{\rho}$ of $\tilde{S}'$ acting on the image $\tilde{k}$ of a lift of k. Since $\mathbf{dev}\circ\tilde{\rho}$ is another projective immersion $\tilde{S}' \to \mathbf{S}^2$, it follows that $\mathbf{dev}\circ\tilde{\rho} = h(\tilde{\rho})\circ\mathbf{dev}$ for an element $h(\tilde{\rho})$ of $\mathrm{Aut}(\mathbf{S}^2)$.

Letting ϑ be the deck transformation of $\tilde{S}'$ corresponding to $\tilde{k}$ and k, we see that ϑ and $\tilde{\rho}$ commute as ρ is a topological automorphism of S'. Hence given a development pair $(\mathbf{dev}, h)$ of Σ', $h(\vartheta)$ and $h(\tilde{\rho})$ act on the convex domain $\mathbf{dev}(S')$ and an open geodesic $\mathbf{dev}(\tilde{k})$. The boundary of $\mathbf{dev}(S')$ is the union of the closure of $\mathbf{dev}(\tilde{k})$ and an open arc α connecting the endpoint of $\mathbf{dev}(\tilde{k})$. Since S' is convex and has a negative Euler characteristic, $h(\vartheta)$ is either hyperbolic or quasi-hyperbolic by Proposition 4.5 of [**12**]. The obvious $h(\vartheta)$-invariant subsets are simply convex triangles with vertices fixed points of $h(\vartheta)$ when $h(\vartheta)$ is hyperbolic and are lunes when $h(\vartheta)$ is quasi-hyperbolic (see Figures 2 and 3 of [**13**]).

The closure of $\mathbf{dev}(S')$ is an $\langle h(\vartheta)\rangle$-invariant set. Since k is strong, we can show that α is a subset of an open $\langle h(\vartheta)\rangle$-invariant triangle or lune as above. Since both $h(\vartheta)$ and $h(\tilde{\rho})$ act on the common arc α in the open invariant triangle or lune, we can verify by explicit calculations that both $h(\vartheta)$ and $h(\tilde{\rho})$ lie in a common 1-parameter subgroup H of $\mathrm{Aut}(\mathbf{S}^3) = \mathrm{SL}_{\pm}(3, \mathbf{R})$.

For any regular neighborhood of $\mathbf{dev}(\tilde{k})$ in $\mathbf{dev}(\Sigma')$, there exists an H-invariant arc $\tilde{\alpha}_k$ connecting the endpoints of $\mathbf{dev}(\tilde{k})$. Now, $\tilde{\alpha}_k$ corresponds to a simple closed arc in α_k freely homotopic to k in Σ'^o. If we choose α_k for each component k, then there exists a compact surface Σ'' in Σ'^o bounded by α_ks; and ρ acts on Σ''. As Σ'' is a compact subsurface in Σ'^o, it is a compact metric space under the Hilbert metric of Σ'^o. Since the group of isometries of a compact metric space is compact, the closure of $\langle\rho\rangle$ is a compact Lie group acting on Σ''. If the closure of $\langle\rho\rangle$ is discrete, then ρ is of finite order. Since the identity component of this Lie group thus does not consist of a point, the Lie group includes the compact Lie subgroup $\mathbf{S}^1$ homeomorphic to a circle acting on Σ'' nontrivially. Since Σ'' is homotopy equivalent to Σ', this is absurd as there is no nontrivial $\mathbf{S}^1$-action on a surface of negative Euler characteristic. Therefore ρ is of finite order. □

PROPOSITION 3.2. *A compact radiant affine n-manifold M with totally geodesic or empty boundary admits a total cross-section Σ to the radial flow if and only if it is affinely diffeomorphic to a generalized affine suspension over a compact real projective $(n-1)$-manifold Σ' with totally geodesic or empty boundary. Moreover, in the above case, Σ with the induced real projective structure from the affine space by the radial flow is real projectively diffeomorphic to Σ'.*

PROOF. Suppose that M has a total cross-section Σ to the radial flow. Then Σ obviously has a real projective structure since the radial flow gives a local projection of the affine space to the real projective space. The affine transition maps preserving the radial flow now become real projective transition maps.

The existence of the total cross-section shows that there exists a cover M' of M homeomorphic to $\Sigma \times \mathbf{R}$. Let $\mathbf{dev} : \tilde{M}' \to \mathbf{R}^n - \{O\}$ for the universal cover $\tilde{M}'$ of M' denote the developing map and $h : \pi_1(M') \to \mathrm{GL}(n, \mathbf{R})$ the holonomy homomorphism, considering $\mathrm{GL}(n, \mathbf{R})$ as the subgroup of $\mathrm{Aut}(\mathcal{H})$ acting on $\mathcal{H}$ fixing O where $\mathcal{H}$ is the n-hemisphere whose interior is identified with $\mathbf{R}^n$.

The inverse image $\tilde{\Sigma}$ of Σ in $\tilde{M}'$ is obviously the universal cover of Σ. $\tilde{M}'$ is obviously homeomorphic to $\tilde{\Sigma} \times \mathbf{R}$, and $\pi_1(\Sigma)$ can be identified with with $\pi_1(M')$ since the deck transformations of M' act on $\tilde{\Sigma}$ as well. The radial projection map $\Pi : \mathbf{R}^n - \{O\} \to \mathbf{S}^{n-1}_\infty$ composed with $\mathbf{dev}|\tilde{\Sigma}$ is the developing map for Σ considered now as a real projective manifold, and $k : \pi_1(\Sigma) \to \mathrm{Aut}(\mathbf{S}^{n-1}_\infty)$, which is obviously obtained from $h : \pi_1(\Sigma) \to \mathrm{GL}(n, \mathbf{R})$ by restriction to $\mathbf{S}^{n-1}_\infty$ is the holonomy homomorphism for Σ.

From the real projective $(n-1)$-manifold Σ, we will now construct an affine structure for $\Sigma \times \mathbf{R}$. We may construct a developing map $\mathbf{dev}' : \tilde{\Sigma} \times \mathbf{R} \to \mathbf{R}^n - \{O\}$ as in the construction of the generalized affine suspension using $\Pi \circ \mathbf{dev}|\tilde{\Sigma}$ by radial extension. Letting h' denote the composition

$$\pi_1(\Sigma \times \mathbf{R}) \xrightarrow{i} \pi_1(\Sigma) \xrightarrow{k} \mathrm{Aut}(\mathbf{S}^{n-1}_\infty),$$

where i is the obvious identification, we choose the lift ρ of h' into $\mathrm{GL}(n, \mathbf{R})$ so that ρ equals above h on $\pi_1(\Sigma)$. Then we define the action of the deck transformation ϑ in $\pi_1(\Sigma \times \mathbf{R})$ on $\tilde{\Sigma} \times \mathbf{R}$ by $\vartheta(x, t) = (i(\vartheta)(x), t + \log ||\rho(\vartheta)(u(x))||)$ where $u(x)$ is the unit vector in $\mathbf{R}^n$ in the direction of $\mathbf{dev}(x)$. Denote by G the group of deck transformations on $\tilde{\Sigma} \times \mathbf{R}$ thus defined.

Since $\mathbf{dev}'$ and $\mathbf{dev}$ are radially complete maps (see Lemma 3.1), it is easy to see that $\mathbf{dev}|\tilde{M}'$ lifts to an affine diffeomorphism $k : \tilde{M}' \to \tilde{\Sigma} \times \mathbf{R}$, so that $\mathbf{dev}' \circ k = \mathbf{dev}$. Then by our choice of ρ and an analytic continuation from a small open set, we see that $k \circ \vartheta = \vartheta' \circ k$ where ϑ and ϑ' are corresponding deck transformations of M'_h and $\Sigma' \times \mathbf{R}$. This means that M' and $(\Sigma' \times \mathbf{R})/G$ are affinely diffeomorphic. Since M is obtained by an action of an infinite cyclic automorphism group from M', transferring the affine action of the cyclic group to $(\tilde{\Sigma} \times \mathbf{R})/G$ shows that M is affinely diffeomorphic to a generalized affine suspension over Σ'.

The converse part is obvious since a generalized affine suspension over a surface Σ using an automorphism γ is a mapping torus and is homeomorphic to a bundle over $\mathbf{S}^1$ with fibers diffeomorphic to Σ with fiber map induced from that of the projection $\Sigma \times \mathbf{R} \to \mathbf{R}$; a fiber is a total cross-section obviously. □

Let us state Barbot's results [**3**] which we will use in this paper, telling us when a radiant affine 3-manifold is a generalized affine suspension. A closed orbit of the radial flow of a radiant affine 3-manifold is of *saddle type* if the differential of the return map associated with the orbit and the local section has two distinct eigenvalues one of which is greater than 1 and the other less than 1.

THEOREM 3.3 (Barbot). *Let M be a closed radiant affine 3-manifold. If one of the following holds, then M admits a total cross-section.*

- *M includes a totally geodesic surface tangent to the radial flow.*
- *M is a convex affine manifold.*
- *M has a closed orbit of the radial flow that is not of saddle type.*

THEOREM 3.4 (Barbot-Choi (Appendix C)). *If a compact radiant affine 3-manifold M has nonempty totally geodesic boundary, and each boundary component is affinely homeomorphic to a quotient of a convex cone in $\mathbf{R}^2 - \{O\}$ or $\mathbf{R}^2 - \{O\}$ itself by a group of affine motions, then M admits a total cross-section.*

A totally geodesic submanifold of an affine 3-manifold carries a natural induced affine structure as 2-manifolds.

Actually, we can dispense with the assumption on boundary components using Barbot's results or from our proof of the Carrière conjecture. But we do not do so since it requires far more work.

We will now begin to discuss about the ideal sets of the Kuiper completions of the holonomy covers of radiant affine manifolds. Since M is a radiant affine manifold, we divide $M_{h\infty}$ into two parts. We denote by $M^{\mathfrak{i}}_{h\infty}$ the set $\mathbf{dev}^{-1}(\mathbf{S}^{n-1}_\infty)$ in $\check{M}_h$, and is, by the following lemma, a subset of $M_{h\infty}$. We let $M^{\mathfrak{f}}_{h\infty} = M_{h\infty} \cap$

$\mathbf{dev}^{-1}(\mathbf{R}^n - \{O\})$. The two are said to be *infinitely* and *finitely* ideal subsets of $\check{M}_h$ respectively.

LEMMA 3.2. *Let A be a subset of $\check{M}_h$ such that $\mathbf{dev}(A)$ is a subset of $\mathbf{S}^{n-1}_\infty$. Then A is a subset of $M_{h\infty}$.*

PROOF. Since $\mathbf{dev}(M_h)$ is the subset of $\mathcal{H}^o$, A is necessarily a subset of $M_{h\infty}$. □

PROPOSITION 3.3. *Assume that $n \geq 3$. If $M^{\mathrm{f}}_{h\infty} = \emptyset$, M is finitely covered by a Hopf manifold or a half-Hopf manifold. In fact, M is a generalized affine suspension of a real projective $(n-1)$-manifold with a universal cover homeomorphic to $\mathbf{S}^{n-1}$ or an $(n-1)$-hemisphere.*

PROOF. Assume that $n \geq 2$. We start with the case when M is a closed manifold. By Lemma 3.1, $\mathbf{dev}(M_h)$ is a subset of $\mathbf{R}^n - \{O\}$. Let $\gamma : I = [0,1] \to \mathbf{R}^n - \{O\}$ be a $\mathbf{d}$-rectifiable path with endpoints $p = \gamma(0)$ and $q = \gamma(1)$, where p belongs to $\mathbf{dev}(M_h)$. Since $M^{\mathrm{f}}_{h\infty}$ is empty, we see that we can lift γ starting from 0 and can continue without any problem. Thus, it follows that $\mathbf{dev} : M_h \to \mathbf{R}^n - \{O\}$ is a covering map.

Now assume that $n \geq 3$. In this case $\mathbf{dev} : M_h \to \mathbf{R}^n - \{O\}$ is a diffeomorphism and $M_h = \tilde{M}$. Since the unit sphere $\mathbf{S}^{n-1}$ is a compact subset of $\mathbf{R}^n - \{O\}$, the set $L = \mathbf{dev}^{-1}(\mathbf{S}^{n-1})$ is a compact subset of M_h. Since the action of deck transformation is properly discontinuous, there exists a deck transformation ϑ such that $\vartheta(L)$ is disjoint from L. Now, $\vartheta(L)$ and L bound a subset of M_h diffeomorphic to $\mathbf{S}^{n-1} \times I$ for an interval I. Clearly $N = M_h/\langle\vartheta\rangle$ is a Hopf manifold and covers M finitely.

Our manifold M is covered by the Hopf manifold N finitely. Obviously there exists a map $f : N \to \mathbf{S}^1$ as N is a mapping torus of $\mathbf{S}^{n-1}$. Regarding $\mathbf{S}^1$ as a quotient space of the real line $\mathbf{R}$, the radiant vector field takes positive values under the closed and non-exact 1-form df on N.

Since the radiant vector field on N is pushed to a radiant vector field on M, a 1-form κ on M obtained by averaging df over the inverse images of evenly covered neighborhood takes positive values for the radiant vector-field. Since κ is an average of closed forms locally, κ is closed. Hence by Tischler [**46**], we see that M admits a total cross-section Σ to the radial flow. (See the proof of Lemma C.3 of C for more details.) Hence M is a generalized affine suspension over Σ by Proposition 3.2. The universal cover of N is homeomorphic to $\mathbf{R}^n - \{O\}$, which is homeomorphic to the product of $(n-1)$-sphere with the real line $\mathbf{R}$. Since the universal cover of N is also homeomorphic to $\tilde{\Sigma} \times \mathbf{R}$ for a cover of $\tilde{\Sigma}$ of Σ, $\tilde{\Sigma}$ is homotopy equivalent to $\mathbf{S}^{n-1}$. Since $\tilde{\Sigma}$ is a total cross-section, it is a closed real projective manifold. It is now obvious that $\tilde{\Sigma}$ is projectively diffeomorphic to $\mathbf{S}^{n-1}$ as a developing map must be such a diffeomorphism.

Suppose now that M has a nonempty boundary. Let S be a boundary component, and let S_h be a component of $p^{-1}(S)$ in M_h. Then as there are no ideal points of M_h, the argument in the first paragraph shows that $\mathbf{dev}|S_h$ maps to $P - \{O\}$ as a covering map where P is a two-dimensional subspace. As $-\mathrm{I}$ commutes with all holonomy elements, and $-\mathrm{I}$ acts on any cover of $P - \{O\}$, we see that we can obtain a double $\hat{M}$ of M which is closed and has a radiant affine structure. Therefore our proposition follows from the first part. □

We will now show that $\check{M}_h \cap \mathbf{dev}^{-1}(O)$ is a unique point in $M_{h\infty}$.

LEMMA 3.3. 1. *We have the inequality*

$$C(t)^{-1}\mathbf{d}(x,y) \le \mathbf{d}(\Phi_{h,t}(x), \Phi_{h,t}(y)) \le C(t)\mathbf{d}(x,y) \tag{3.4}$$

for each $x, y \in M_h$ and a positive number $C(t)$ depending only on t smoothly.

2. *Each $\Phi_{h,t}$ extends to a homeomorphism $\Psi_{h,t}$ of $\check{M}_h$ for $t \in \mathbf{R}$, and the above inequality holds for $\Psi_{h,t}$ also.*
3. *Obviously, $\Psi_{h,t}$ acts on M_h and hence on $M_{h\infty}$ and on $M^{\mathrm{f}}_{h\infty}$ and fixes each point of $M^{\mathrm{i}}_{h\infty}$.*

PROOF. (1) For every pair of points x and y, $\mathbf{d}(x,y)$ is the infimum of $\mathbf{d}$-lengths of paths α connecting x and y on M_h. For each path α connecting x and y, $\Phi_{h,t}\circ\alpha$ connects $\Phi_{h,t}(x)$ and $\Phi_{h,t}(y)$. For the positive number $C(t)$ from equation 3.1, the $\mathbf{d}$-length of $\Phi_{h,t}\circ\alpha$ is bounded below by $C(t)^{-1}$ times that of α and bounded above by $C(t)$ times that of α. Hence, we get the inequality.

(2) Since by (1), $\Phi_{h,t}$ is a quasi-isometry $M_h \to M_h$, $\Phi_{h,t}$ extends to a continuous map $\Psi_{h,t} : \check{M}_h \to \check{M}_h$, where obviously the inverse of $\Psi_{h,t}$ is $\Psi_{h,-t}$.

(3) Since $\Psi_{h,t}$ acts on M_h, $\Psi_{h,t}$ acts on its complement $M_{h\infty}$. Since $\Psi_{h,t}$ is a continuous extension of $\Phi_{h,t}$, the following diagram also commutes

$$\begin{array}{ccc} \check{M}_h & \xrightarrow{\Psi_{h,t}} & \check{M}_h \\ \mathbf{dev}\downarrow & & \downarrow\mathbf{dev} \\ \mathcal{H} & \xrightarrow{\Phi'_t} & \mathcal{H}. \end{array} \tag{3.5}$$

Since Φ'_t acts on $\mathbf{R}^n - \{O\}$, it follows that $\Psi_{h,t}$ acts on $M^{\mathrm{f}}_{h\infty}$.

For each point x of $M^{\mathrm{i}}_{h\infty}$, there exists a sequence $x_i \in M_h$ converging to x, where a radial line l_i passing through x_i for each i. l_i has two endpoints p_i and q_i so that $\mathbf{dev}(p_i) \in \mathbf{S}^{n-1}_\infty$ and $\mathbf{dev}(q_i) = O$ since $\mathbf{dev}|l_i$ is a radial line connecting O and a point of $\mathbf{S}^{n-1}_\infty$. Since l_i is a flow line of $\mathbf{V}_h$, $\Psi_{h,t}$ acts on l_i and hence on its closure $\mathrm{Cl}(l_i)$; thus, $\Psi_{h,t}$ fixes each p_i. Since p_i converges to x as well, and $\check{M}_h$ is a metric space, x is fixed by $\Psi_{h,t}$ for each $t \in \mathbf{R}$. □

REMARK 3.2. By an abuse of notation, we will denote $\Psi_{h,t}$ by $\Phi_{h,t}$ from now on.

DEFINITION 3.1. A *radial* segment in $\check{M}_h$ is a convex segment in $\check{M}_h$ such that the image of its endpoint consists of the origin and a point of $\mathbf{S}^{n-1}_\infty$.

One sees that the closure of any radial line is a radial segment.

PROPOSITION 3.4. *$\check{M}_h$ includes a point p of a radial segment in $\check{M}_h$ satisfying $\mathbf{dev}(p) = O$. Furthermore, given a point q of $\check{M}_h$ such that $\mathbf{dev}(q) = O$, then $p = q$.*

PROOF. An endpoint p of a radial maximal line l_1 clearly satisfies $\mathbf{dev}(p) = O$ by Lemma 3.1.

Since q belongs to $\check{M}_h$, there exists a path α with endpoint q and a point x of M_h. Choose a path β from a point y of l_1 ending at x and produce a path γ from y to q obtained by joining α and β at y. Since $\mathbf{dev}\circ\gamma$ is an arc in $\mathbf{R}^n$ bounded away from $\delta\mathcal{H}$, $\mathbf{dev}\circ\Phi_{h,t}\circ\gamma$ is a path connecting $\mathbf{dev}\circ\Phi_{h,t}(y)$ and $\mathbf{dev}(q)$, and has $\mathbf{d}$-length less than or equal to a constant $c(t)$ times that of γ. As $t \to -\infty$, it is easy to see that $c(t) \to 0$. Hence as $t \to -\infty$, the $\mathbf{d}$-length of $\Phi_{h,t} \circ \gamma$ goes to zero, meaning that $\Phi_{h,t}(y)$ converges to q. Clearly, $\Phi_{h,t}(y)$ converges to p since $\Phi_{h,t}(y)$ belongs to l_1; thus, we obtain $p = q$. □

We denote the unique point p such that $\mathbf{dev}(p) = O$ by O, and say it is the *origin* of $\check{M}_h$ or M_h; O is fixed by every deck transformation of M_h.

LEMMA 3.4. *Through each point of $\check{M}_h$ a radial segment passes. Every radial segment l is either a subset of $M_{h\infty}$, or l with its endpoints removed belongs to M_h. The endpoints of l consists of O and a point of $M^{\mathfrak{i}}_{h\infty}$.*

PROOF. Let x be a point of $\check{M}_h$. If $x = O$, then we are done by Proposition 3.4. Assume $x \neq O$, which means that $\mathbf{dev}(x) \neq O$, and there exists a sequence $\{x_i\}$ of points of M_h converging to x with respect to $\mathbf{d}$ such that $\mathbf{d}(O, \mathbf{dev}(x_i)) \geq C$ for some $C > 0$. Since $x_i \in M_h$, there exists a radial segment l_i containing x_i such that $l_i^o \subset M_h$. Then since $x_i \to x$ with respect to $\mathbf{d}$, there exists a path α_{ij} in M_h with endpoints x_i and x_j so that for each $\epsilon > 0$, there exists N such that if $i, j > N$, then the $\mathbf{d}$-length(α_{ij}) are less than ϵ.

Let $\mathbf{S}^2_1$ denote the $\mathbf{d}$-sphere of radius $\pi/4$ with center O. For each i, let y_i denote the $\mathbf{d}$-midpoint of l_i, and Π denote the radial projection of $\mathcal{H} - \{O\}$ to $\mathbf{S}^2_1$, the radial completeness of M_h shows that $\Pi \circ \mathbf{dev} \circ \alpha_{ij}$ lifts to a path β_{ij} in M_h connecting y_i and y_j. Since $\mathbf{dev}(x_i)$ is uniformly bounded away from O, it follows that the $\mathbf{d}$-length of β_{ij} is bounded above by $c(C)$ times that of α_{ij}, where $c(C)$ is a positive constant depending only on C easily calculable from spherical geometry. Hence y_i converges to y in $\check{M}_h$ and l_i converges to a radial segment l which contains x and y.

Let l be an arbitrary radial segment. Then $\Phi_{h,t}$ for $t \in \mathbf{R}$ acts transitively on l^o. Since $\Phi_{h,t}$ acts on M_h and on $M_{h\infty}$, it follows that either $l^o \subset M_h$ or l is a subset of $M_{h\infty}$. □

DEFINITION 3.2. A *radiant* set in M_h is a $\Phi_{h,t}$-invariant subset of M_h, and a *radiant* set in $\check{M}_h$ is a $\Phi_{h,t}$-invariant closed subset of $\check{M}_h$ that is not a subset of $M^{\mathfrak{i}}_{h\infty} \cup \{O\}$.

Obviously, a radiant set in $\check{M}_h$ is a union of maximal radial segments. A *radiant* set in $\mathcal{H}$ is a Φ'_t-invariant subset of M_h that is not a subset of $\delta\mathcal{H} \cup \{O\}$. A closed radiant set in $\mathcal{H}$ is a union of radial segments.

LEMMA 3.5. *Let S be a closed radiant subset of $\check{M}_h$. Then $S \cap M_h$ and $S \cap M^{\mathfrak{f}}_{h\infty}$ are closed radiant sets also.*

EXAMPLE 3.2. Let N be a Benzécri suspension of a closed real projective surface Σ. Then N_h may be identified with $\Sigma_h \times (0, 1)$, where Σ_h is a holonomy cover of Σ. Then the Kuiper completion of N_h can be easily identified with $(\check{\Sigma}_h \times [0, 1])/\sim$ where $\sim$ is the equivalence relation identifying all points of $\check{\Sigma}_h \times 0$ with one point. The previous results in this section are easily verified for this example.

A *radiant* n-bihedron L is an n-bihedron in $\check{M}$ such that the boundary of $\mathbf{dev}(L)$ contains O and $\mathbf{dev}(F)$ for a side F of L lies in $\mathbf{S}^{n-1}_\infty$, which is obviously a radiant set. We define α_L to be the interior of F.

LEMMA 3.6. *Suppose that M is not convex. Then the boundary δB of a convex compact n-ball B in $\check{M}_h$ (resp. $\check{M}$) meets M_h (resp. $\tilde{M}$). Furthermore, a radiant*

n-bihedron is a bihedral n-crescent, and $\check{M}_h$ (resp. $\check{M}$) cannot contain any n-hemisphere.

PROOF. If δB does not meet M_h, then M_h is a subset of B, $\delta B \subset M_{h\infty}$, and M_h is boundaryless. Therefore, we obtain $M_h = B^o$, which is a contradiction since B^o is convex.

Let L be a radiant n-bihedron. Since the side F of L with $\mathbf{dev}(F) \subset \mathbf{S}^2_\infty$ is a subset of $M^{\mathrm{i}}_{h\infty}$, and by the above statement, the intersection of the other side of L with M_h is not empty, it follows that L is a bihedral n-crescent.

If $\check{M}_h$ includes an n-hemisphere L, then $\mathbf{dev}|L : L \to \mathcal{H}$ is an imbedding onto $\mathcal{H}$. Since $\mathbf{dev}(M_h)$ is a subset of $\mathcal{H}^o$, δL does not meet M_h. The first part of this lemma says that this cannot happen. □

PROPOSITION 3.5. *Let M be a compact radiant affine manifold ($n \geq 2$). If M is not $(n-1)$-convex, then $\check{M}_h$ includes a radiant n-bihedron.*

PROOF. By Theorem 4.6 of [**15**], $\check{M}_h$ includes an n-crescent L. L is a bihedral n-crescent by Lemma 3.6.

By Lemma 3.1, $\mathbf{dev}(L)^o$ is disjoint from O. Let L' be the union of $\Phi_{h,t}(L^o)$ for all $t \in \mathbf{R}$, which is an open subset of $\tilde{M}$. Since by equation 3.2, $\mathbf{dev}(L')$ is the union of $\Phi'_{h,t}(\mathbf{dev}(L^o))$ for all t, $\mathbf{dev}(L')$ is a radiant open half-space in $\mathcal{H}^o$. For any t and t' such that $\Phi_{h,t}(L^o)$ and $\Phi_{h,t'}(L^o)$ are overlapping, $\mathbf{dev}|\Phi_{h,t}(L^o) \cup \Phi_{h,t'}(L^o)$ is a diffeomorphism onto $\mathbf{dev}(\Phi_{h,t}(L^o)) \cup \mathbf{dev}(\Phi_{h,t'}(L^o))$ by Remark 1.3. By choosing a sequence $\{t_i\}$ such that $\Phi_{h,t_i}(L^o)$ and $\Phi_{h,t_{i+1}}(L^o)$ always overlap, and by an induction, we see that $\mathbf{dev}|L'$ is a diffeomorphism to an open half-space. For the closure L'' of L' in $\check{M}$, $\mathbf{dev}|L''$ is an imbedding onto the closure of the half-space in $\mathcal{H}$ (see Chapter 1). Hence, L'' is a radiant n-bihedron. □

PROPOSITION 3.6. *Suppose that M is not convex and M_h includes an open n-bihedron. Then $\check{M}_h$ includes a unique radiant n-bihedron including the open bihedron. In fact, given any n-crescent, there exists a unique radiant n-bihedron including it.*

PROOF. The existence part is analogous to the proof of the above theorem; that is, we radially extend it. The uniqueness follows from Proposition 1.1. □

Finally, we say about the intersection properties of radiant n-bihedra (see Appendix A for definition of transversal intersection):

COROLLARY 3.1. *Suppose that M is not convex. Let R_1 and R_2 be two overlapping radiant n-bihedra in $\check{M}_h$. Then $R_1 = R_2$ or R_1 and R_2 as n-crescents intersect transversally.*

PROOF. By Theorem A.1 we need only to show that R_1 cannot be a proper subset of R_2 and vice versa. This is so since $\mathbf{dev}(R_1)$ being radiant cannot be a proper subset of $\mathbf{dev}(R_2)$ from elementary geometry. □

CHAPTER 4

Three-dimensional radiant affine manifolds and concave affine manifolds

We will now assume that M is a three-dimensional compact radiant affine manifold with totally geodesic or empty boundary and is not two-convex from now on. Since $\check{M}_h$ does not contain any hemispheric n-crescents by Lemma 3.6, Corollary 1.2 of [**15**] shows that after splitting along the two-faced submanifolds arising from bihedral n-crescents, the resulting manifold N decomposes into concave affine manifolds with concave boundary and two-convex affine manifolds with convex boundary. The concave boundary may not be totally geodesic. However we will show in the next chapter that the concave boundary of the concave affine manifold is totally geodesic in the radiant affine 3-manifold; our decomposition takes place along totally geodesic closed surfaces in M. (For full treatment of the decomposition theory, see [**15**].)

We suppose that $\check{M}$ includes a bihedral 3-crescent since otherwise M will be 2-convex by [**15**] and we can proceed to the next chapter. By the previous chapter, any bihedral 3-crescent is a subset of a unique radiant 3-bihedron. Let B denote the set of all bihedral 3-crescents and $\mathcal{B}$ that of all radiant 3-bihedra. We recall the equivalence relation on B defined in [**15**]. Given radiant 3-crescents R and S, we say R and S are simply equivalent if they overlap, which generates an equivalence relation $\sim$ on the set of all bihedral 3-crescents. So we define as in [**15**] for a bihedral 3-crescent R:

$$\Lambda(R) = \bigcup_{S \sim R} S, \quad \delta_\infty \Lambda(R) = \bigcup_{S \sim R} \alpha_S, \quad \Lambda_1(R) = \bigcup_{S \sim R} S - \nu_S.$$

Given radiant 3-bihedra R and S, we say that R and S are simply equivalent if they overlap, which again generates an equivalence relation $\sim'$ on $\mathcal{B}$. We define for a radiant bihedron R:

$$\Lambda'(R) = \bigcup_{S \sim' R} S, \quad \delta_\infty \Lambda'(R) = \bigcup_{S \sim' R} \alpha_S, \quad \Lambda'_1(R) = \bigcup_{S \sim' R} S - \nu_S.$$

We easily see that given a 3-crescent T, $\Lambda'(T') \subset \Lambda(T)$ for the unique radiant 3-bihedron T' including T since any radiant bihedron equivalent to T' is equivalent to T; conversely, by Proposition 3.6, we have $\Lambda'(T') = \Lambda(T)$. Since radiant 3-bihedra are radiant, $\Lambda'(R)$ is a radiant set; $\Lambda'(R) \cap M_h$ and $\Lambda(R) \cap M_h$ are radiant sets.

Since any 3-crescent R is included in a unique radiant bihedron R', a radiant bihedron is a 3-crescent by Lemma 3.6, and α_R being in $M_{h\infty}$ cannot meet R'^o, it follows that $\alpha_R = \alpha_{R'}$ with $\alpha_R \subset M^{\mathsf{i}}_{h\infty}$. Hence, $\delta_\infty \Lambda(R) = \delta_\infty \Lambda'(R')$ and $\Lambda_1(R) = \Lambda'_1(R')$. (Because of these properties, we will drop primes from $\Lambda'(R)$ and $\sim'$, $\delta_\infty \Lambda'(R)$, and $\Lambda'_1(R)$ respectively for a given radiant bihedron R.)

Let us recall properties of $\Lambda(R), \Lambda_1(R)$, and $\delta_\infty \Lambda(R)$ from Chapter 7 of [**15**], the proofs of which are given there. $\Lambda(R)$ and $\Lambda_1(R)$ are path connected. Given a natural projective structure induced from 3-crescents, $\Lambda_1(R)$ is a real projective 3-manifold with boundary $\delta_\infty\Lambda(R)$ a totally geodesic open surface. Since Corollary 3.1 shows that for two overlapping 3-crescents R_1 and R_2, α_{R_1} and α_{R_2} extend each other into a larger surface, there exists a unique great sphere $\mathbf{S}^2$ including $\mathbf{dev}(\delta_\infty\Lambda(R))$ and a unique component A_R of $\mathbf{S}^3 - \mathbf{S}^2$ such that $\mathbf{dev}(\Lambda(R)) \subset \mathrm{Cl}(A_R)$ and $\mathbf{dev}(\Lambda(R) - \mathrm{Cl}(\delta_\infty\Lambda(R))) \subset A_R$. In our case $\mathbf{S}^2$ must equal $\mathbf{S}^2_\infty$ and A_R equal $\mathcal{H}^o$ by our fixed choice. For a deck transformation ϑ acting on $\Lambda(R)$, A_R is $h(\vartheta)$-invariant. $\Lambda(R)\cap M_h$ is a closed subset of M_h. The topological boundary $\mathrm{bd}\Lambda(R)\cap M_h$ is a properly imbedded topological surface in M_h^o, and $\Lambda(R)\cap M_h$ is a topological submanifold of M_h with concave boundary $\mathrm{bd}\Lambda(R)\cap M_h$.

Given a deck transformation ϑ, we have $\vartheta(\Lambda(R)) = \Lambda(\vartheta(R))$, $\vartheta(\Lambda_1(R)) = \Lambda_1(\vartheta(R))$, and $\vartheta(\delta_\infty\Lambda(R)) = \delta_\infty\Lambda(\vartheta(R))$.

PROPOSITION 4.1. *Given two radiant bihedra R and S, precisely one of the following possibilities holds:*

- $\Lambda(R) = \Lambda(S)$ *and* $R \sim S$.
- $\Lambda(R)\cap M_h$ *and* $\Lambda(S)\cap M_h$ *are disjoint and* $R \not\sim S$.
- $\Lambda(R)\cap\Lambda(R)\cap M_h$ *equals the union of common components of* $\mathrm{bd}\Lambda(R)\cap M_h$ *and* $\mathrm{bd}\Lambda(S)\cap M_h$ *where* $R\not\sim S$, *which furthermore equals the union of common components of* $\nu_T\cap M_h$ *and* $\nu_U\cap M_h$ *for two radiant bihedra* $T\sim R$ *and* $U\sim S$.

We recall from [**15**] that the *copied* components arising from bihedral 3-crescents are common components of $\mathrm{bd}\Lambda(R)\cap M_h$ and $\mathrm{bd}\Lambda(T)\cap M_h$ for some pair of inequivalent radiant bihedra R and T.

LEMMA 4.1. *The collection whose elements are copied components are locally finite, the union A of all copied components is a properly imbedded submanifold in M_h, and $p|A : A \to p(A)$ covers a compact submanifold $p(A)$ of codimension one.*

A is said to be the *pre-two-faced submanifold*, and $p(A)$ the *two-faced submanifold.* If C is a copied component of $\nu_{R'}\cap M_h$ in A, then $p(C)$ is a component of the closed codimension-one submanifold $p(A)$. Since for a radiant 3-bihedron R', $\nu_{R'}\cap M_h$ is foliated by radial lines, the following lemma shows that $p(A)$ consists of components that are totally geodesic two-sided tori or one- or two-sided Klein bottles.

LEMMA 4.2. *Suppose that a closed projective surface S is covered by an open subset C of $\mathbf{R}^2-\{O\}$ foliated by radial lines. Then S is an affine torus, or an affine Klein bottle covered by a convex open cone in $\mathbf{R}^2-\{O\}$ or covered by $\mathbf{R}^2-\{O\}$ itself.*

PROOF. Suppose that C is a proper subset of $\mathbf{R}^2-\{O\}$. Given on $\mathbf{R}^2$ polar coordinates (r,θ) where $r,\theta\in\mathbf{R}$, C is given by $\theta_1<\theta<\theta_2$ where $\theta_2-\theta_1<2\pi$. Suppose that $\theta_2-\theta_1>\pi$. Let l_1 and l_2 be the radial lines corresponding to θ_1 and θ_2 respectively. Then a radial line m_1 in the opposite direction to l_1 and m_2 to l_2 both lie in C.

Since $p(C)$ is a closed surface, a subgroup G of $\pi_1(M)/\pi_1(M_h)$ acts on C so that p induces a homeomorphism $C/G\to p(C)$. Since C is invariant under the affine action of G, it follows that $m_1\cup m_2$ is G-invariant, and $(m_1\cup m_2)/G$ is a simple closed curve or the union of two simple closed curves. As in Lemma 2 of [**11**], this means that a double cover of the surface $p(C)$ is homotopy equivalent to a simple closed curve in the cover, and such a closed surface does not exist. Hence, $\theta_2 - \theta_1 \leq \pi$, and C is convex. Since G preserves the foliation, S is foliated; S is homeomorphic to a torus or a Klein bottle.

If C equals $\mathbf{R}^2 - \{O\}$, then S is again foliated by radial lines, and is homeomorphic to a torus or a Klein bottle. □

We split M and M_h by $p(A)$ and A respectively, and we obtain an n-manifold N and its cover M' including a copy of $M_h - A$ as a dense subset such that the covering map $p|M_h - A : M_h - A \to M - p(A)$ naturally extends to a covering $p : M' \to N$ and $\mathbf{dev}|M_h - A$ extends to a developing map $\mathbf{dev} : M' \to \mathbf{S}^3$. As explained in Chapter 8 of [**15**], given a component N_i of N, we choose a component P_i of $M_h - A$ covering a component of $M - A$ dense in N_i. Then the disjoint union of P_i is a holonomy cover of $M - A$, and the disjoint union of L_i where L_i is the component of M' where P_i is dense in is a holonomy cover of N. (Note that the cover is not necessarily connected here.) We denote the disjoint union by N_h.

Since M' admits an immersion to $\mathbf{S}^3$, we may pull-back the standard metric μ to M' and obtain a completion $\check{M}'$ of M'. We can show that the completion $\check{M}'$ of M' includes radiant bihedra with same interiors as those of $\check{M}_h$; there is a one-to-one correspondence of radiant bihedra in this way. Using this we can show easily that in $\check{M}'$, there are no more copied components of $\mathrm{bd}\Lambda(R) \cap N_h$ for every crescent R in $\check{M}'$. (See Chapter 8 of [**15**] for details.) Hence, in $\check{N}_h$ there are no copied components.

Let $\mathcal{A} = \bigcup_{R \in B'} \Lambda(R) \cap N_h$ where B' is the set of representatives of the equivalence classes of radiant bihedra in $\check{N}_h$ under $\sim$. Since we did the splitting, given two radiant bihedra R and S, either $\Lambda(R) = \Lambda(S)$ or $\Lambda(R) \cap M_h$ and $\Lambda(S) \cap M_h$ are disjoint. Since deck transformations send radiant bihedra to radiant bihedra, the deck transformation group acts on $\mathcal{A}$. Since the collection consisting of elements of form $\Lambda(R')$ is locally finite (see [**15**]), $\mathcal{A} \cap N_h$ is a submanifold of N_h. It follows that $\mathcal{A}$ covers a compact codimension 0 submanifold L of N with boundary δL in N^o by Proposition 2.1; L is turns out to be a concave affine manifold of type II. (See Chapters 7 and 10 of [**15**] for more details.)

Proposition 4.2 shows that $\mathrm{bd}L$ is totally geodesic since each component of $\mathrm{bd}L$ is covered by a component of $\mathrm{bd}\Lambda(R) \cap N_h$ for some R.

PROPOSITION 4.2. *$\mathrm{bd}\Lambda(R) \cap N_h$ is a totally geodesic submanifold in N_h^o.*

PROOF. See Chapter 5 for the proof. □

Since $\Lambda(R) \cap N_h$ is radiant, $\mathrm{bd}\Lambda(R) \cap N_h$ is also foliated by radial lines.

PROPOSITION 4.3. *Each component of $\mathrm{bd}\Lambda(R) \cap N_h$ is a component of $\nu_T \cap N_h$ for a radiant bihedron T, $T \sim R$.*

PROOF. Let x be a point of a component F of $\mathrm{bd}\Lambda(R) \cap N_h$ so that $x \in T$ for $T \sim R$. Since x is a boundary point, $x \in \nu_T$. The totally geodesic surface F and the component F' of $\nu_T \cap N_h$ containing x are tangent at x since otherwise, F intersects T^o which is absurd. Since F and F' are both totally geodesic and properly imbedded, we obtain $F = F'$. □

THEOREM 4.1. *A compact radiant affine 3-manifold M decomposes into two-convex radiant affine manifolds and radiant concave affine manifolds of type II along totally geodesic affine tori or Klein bottles covered by convex open cones in $\mathbf{R}^2$ or* *covered by* $\mathbf{R}^2 - \{O\}$. *The Kuiper completion of the holonomy cover of each of the two-convex radiant affine manifolds includes no* 3*-crescents or radiant bihedra.*

CHAPTER 5

The decomposition along totally geodesic surfaces

The aim of this chapter is to prove Proposition 4.2 needed in Chapter 4. We will use the notations of the chapter: We prove that $\mathrm{bd}\Lambda(R)\cap N_h$ is totally geodesic by showing that at each tiny ball neighborhood $B(x)$ of $x \in \mathrm{bd}\Lambda(R)\cap N_h$, there exists a unique supporting plane to the convex ball $B(x) - \Lambda(R)$. If this is not true, we will show that there exist two transversally intersecting crescents nearby x, which will be shown to be a contradiction to certain maximal properties of $\Lambda(R)$ by Lemma 5.1.

We begin the proof. Let $x \in \mathrm{bd}\Lambda(R)\cap N_h$ and the tiny ball neighborhood $B(x)$ of x. Then since $x \in N_h^o$, $B(x)^o$ is an open neighborhood of x and $B(x)^o - \Lambda(R)$ is a convex open set K by definition of concave boundary (see Chapter 3 of [**15**]). Note that the boundary of K in $B(x)^o$ equals $\mathrm{bd}\Lambda(R)\cap B(x)^o$. Since $\Lambda(R)\cap N_h$ is a radiant set, K and $\mathrm{bd}K$ are foliated by maximal radial lines in $B(x)^o$.

We claim that every point of $\mathrm{bd}\Lambda(R)\cap B(x)^o$ has an identical supporting hyperplane in $B(x)^o$ to K. Then $\mathrm{bd}\Lambda(R)\cap B(x)^o$ equals this hyperplane and is totally geodesic, proving the proposition.

Suppose that the claim is not true –(*). Then we will show that there exist two radiant 3-bihedra T and T' ($T, T' \sim R$) meeting transversally, satisfying $T^o \cap T'^o \cap B(x)^o \neq \emptyset$, $\nu_T \cap \nu_{T'}$ meets $B(x)^o$, and ν_T contains a point of $B(x)^o \cap \mathrm{bd}\Lambda(R)$; our claim will be proved by showing that this cannot happen by Lemma 5.1.

Firstly, suppose that at a point y of $\mathrm{bd}\Lambda(R)\cap B(x)^o$ there are at least two distinct supporting hyperplanes P_1 and P_2 in $B(x)^o$ to K. Since K is foliated by maximal radial lines in $B(x)^o$, $\mathbf{dev}(P_1)$ and $\mathbf{dev}(P_2)$ respectively are included in hyperplanes in $\mathbf{R}^3$ passing through O. Since P_1 and P_2 lie in the closure of the complement of K, it follows that $P_1, P_2 \subset \Lambda(R)\cap B(x)^o$. Since $\Lambda(R)$ is a union of radiant 3-bihedra, it follows that $\Lambda(R)\cap N_h$ is foliated by radial lines, P_1 and P_2 must be foliated by subarcs of radial lines in $B(x)^o$, and they meet at a subarc of a radial line in $B(x)^o$ passing through y.

Let $\mathbf{S}^2_r$ be the sphere with center O and radius $r = \mathbf{d}(O, \mathbf{dev}(x))$. Then $\mathbf{dev}^{-1}(\mathbf{S}^2_r)\cap B(x)^o$ is a smooth hypersurface Σ_x in $B(x)^o$ μ-orthogonal to radial lines.

It is obvious that Σ_x has an induced projective structure from the imbedding $\mathbf{S}^2_r \hookrightarrow \mathbf{S}^3$, i.e., Σ_x has a flat projective structure of an open subset of the sphere $\mathbf{S}^2_r$ with the projective structure induced from its Riemannian metric. Let s_1 and s_2 be arcs in $P_1\cap\Sigma_x$ and $P_2\cap\Sigma_x$ respectively ending at y, oriented away from y, and at points of $\mathrm{bd}B(x)^o$ as shown in Figure 5.1(a). $\Sigma_x \cap K$ is a convex open subset with a boundary point y. The boundary α of $\Sigma_x\cap K$ in Σ_x forms an open arc, and $\alpha - \{y\}$ has two components α_1 and α_2, which we orient away from y. We may assume that s_1 are tangent to α_1 and s_2 to α_2. (See Figure 5.1 (a).)

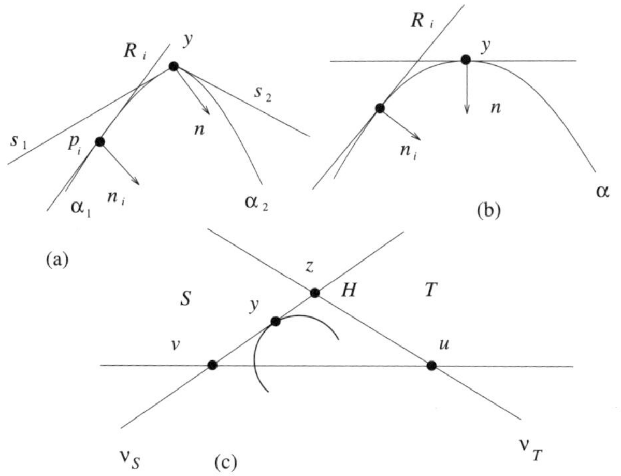

FIGURE 5.1. The cross-section Σ_x of $B(x)$ meeting with crescents.

Choose a sequence $\{p_i\}$ of points of α_1 converging to y. Since α_1 is a subset of $\Lambda(R)$, $p_i \in R_i$ for a radiant 3-bihedron R_i, $R_i \sim R$; as $p_i \in \mathrm{bd}\Lambda(R)$, it follows that $p_i \in \nu_{R_i}$ for each i. Let n_i be the outer normal vector to ν_{R_i} at p_i with respect to the spherical metric $\mathbf{d}$, which is a vector tangent to the spherical surface Σ_x at p_i. Recall that $\nu_{R_i} \cap B(x)$ is a compact $(n-1)$-ball with boundary in $\mathrm{bd}B(x)$ and $R_i^o \cap B(x)$ is a component of the complement of this disk (see Lemma 3.13 of [**15**]); since the endpoint y of s_1 is not a subset of $R_i^o \cap B(x)^o$, it follows easily that the angle θ_i (i.e. the spherical angle) between the tangent vector to the segment on Σ_x connecting y and p_i oriented towards p_i at p_i and n_i is greater than or equal to $\pi/2$ by a geometric consideration (see Figure 5.1). Choosing a subsequence if necessary, we may assume without loss of generality that n_i converges to a unit vector n at y. By Corollary 3.16 of [**15**] (or Section 6.2 of [**11**]), we see that there exists a radiant 3-bihedron T in $\check{M}_h$ with $y \in \nu_T$, n the outer normal vector to T, and $R \sim T$.

Since n is a limit of n_i, the angle between n and the tangent vector to s_1 at y is greater than or equal to $\pi/2$. Since $T \cap B(x)$ is a closure of a component of $T - P$ for an $(n-1)$-ball $P = \nu_T \cap B(x)$, and α_1 is in $\mathrm{bd}\Lambda(R) \cap B(x)^o$, s_1 cannot point towards T^o; otherwise, a small open arc in α_1 near y is included in T^o. Therefore, s_1 and n are perpendicular and $s_1 \subset \nu_T$.

Similarly, we have $s_2 \subset T'$ for a radiant 3-bihedron with $y \in \nu_{T'}$ and a normal vector n' perpendicular to the tangent vector to s_2 at y.

Since s_1 and s_2 make an angle less than π and greater than 0, it follows that n and n' make an angle less than π and greater than 0. Since ν_T and $\nu_{T'}$ both pass y, it follows that $\nu_T \cap \nu_{T'}$ meets $B(x)^o$, and $T^o \cap T'^o \cap B(x)^o \neq \emptyset$, and $T \sim T'$. Since the normal vectors are not parallel, it follows that T and T' intersect transversally by Theorem A.1 and obviously both T and T' contain y in $B(x)^o \cap \mathrm{bd}\Lambda(R)$.

Secondly, suppose now that at every point y of $\mathrm{bd}\Lambda(R) \cap B(x)^o$, there exists a unique supporting hyperplane to K in $B(x)^o$ at y, and $\Sigma_x \cap \mathrm{bd}\Lambda(R) \cap B(x)^o$ is a differentiable convex arc, say α. Since by our assumption (*) α is not projectively geodesic in the sphere Σ_x with the induced projective structure, there is a point y in α with a compact neighborhood in α where there exists an infinite collection of distinct supporting hyperplanes to points of the neighborhood. Hence, we can choose a sequence $\{p_i\}$ of points of $B(x)^o \cap \mathrm{bd}\Lambda(R)$ converging to y of $B(x)^o \cap \mathrm{bd}\Lambda(R)$ so that supporting hyperplanes at p_i are mutually distinct.

We have $p_i \in R_i$ for $R_i \sim R$ for each i. Since p_i does not belong to R_i^o, we have $p_i \in \nu_{R_i}$ for $R_i \sim R$. Since $\nu_{R_i} \subset \Lambda(R)$, it follows that $\nu_{R_i} \cap B(x)^o$ is the unique supporting hyperplane at p_i; thus, R_i are mutually distinct. Similarly to the above argument, y is a point of a radiant 3-bihedron S. Let n_i denote the normal vector to R_i at x_i and n that to S at y. By the uniqueness of tangent lines, it is easy to see that n_i converges to n. This, the convexity of α, and an elementary spherical geometry show that for i sufficiently large $\nu_S \cap \nu_{R_i}$ meets $B(x)^o$, and we have $S^o \cap R_i^o \cap B(x)^o \neq \emptyset$, and $S \not\subset R_i$ and $R_i \not\subset S$ (see Figure 5.1 (b)). Since S and R_i overlap, S and R_i intersect transversally, by Corollary 3.1 and R_i contains p_i in $B(x)^o \cap \mathrm{bd}\Lambda(R)$. Therefore, our proof is completed by the following lemma. □

LEMMA 5.1. *We have two radiant 3-bihedra S and T equivalent to R and overlapping with a tiny ball $B(x)$ such that $\nu_S \cap \nu_T$ meets $B(x)^o$, and S and T meet transversally, and $S^o \cap T^o \cap B(x)^o \neq \emptyset$. Then S cannot contain a point of $B(x)^o \cap \mathrm{bd}\Lambda(R)$,*

PROOF. Suppose not. Let y be a point of ν_S in $B(x)^o \cap \mathrm{bd}\Lambda(R)$. Since S and T meet transversally, $\nu_S \cap \nu_T$ is a segment H with interior in ν_S^o and ν_T^o and boundary in $\delta\nu_S$ and $\delta\nu_T$.

If y does not belong to H, choose a point z of $H \cap \Sigma_x^o$, a point u of $\nu_T \cap \Sigma_x$ nearby z, and v of $\nu_S \cap \Sigma_x$ near y so that the projective geodesic $\overline{zv}$ on Σ_x contains y in the interior. We obtain the triangle $\triangle(vzu)$ in Σ_x with edges $\overline{vz}$ and $\overline{zu}$ lying in ν_S and ν_T respectively. If y belongs to H, then choose a point z to be y and a point u of $\nu_T \cap \Sigma_x$ nearby z, and a point v of $\nu_S \cap \Sigma_x$ near z so that $\triangle(vzu)$ lies in Σ_x and its edges $\overline{vz}$ and $\overline{zu}$ in ν_S and ν_T respectively. By our construction, $\overline{vu}$ is not a subset of $(S \cup T) \cap B(x)^o$ (see Figure 5.1 (c)).

We denote by P the subspace of M_h that is the union of all radial lines passing through $\triangle(vzu)$. Again by equation 3.2, we see that $\mathbf{dev}|P$ is an imbedding onto a convex subset bounded by totally geodesic surfaces in $\mathcal{H}^o$. In particular, $\mathrm{Cl}(P)$ is a radiant tetrahedron in $\check{M}_h$ with a vertex O bounded by three sides F_1, F_2, and F_3 corresponding to $\overline{vu}$, $\overline{vz}$, and $\overline{zu}$ respectively and the side F_4 in $M_{h\infty}^{\mathrm{i}}$.

Since S and T are overlapping, $\mathbf{dev}|S \cup T$ is an imbedding onto $\mathbf{dev}(S) \cup \mathbf{dev}(T)$ where $\mathbf{dev}(S)$ and $\mathbf{dev}(T)$ are two radiant 3-bihedra with sides in $\mathbf{S}^2_\infty$ and $\mathbf{dev}(\nu_S)$ and $\mathbf{dev}(\nu_T)$ meeting transversally. Let H' be the $(n-1)$-hemisphere in $\mathcal{H}$ including F_1. Then H' is a side of a unique radiant 3-bihedron U in $\mathcal{H}$ including $\mathbf{dev}(\mathrm{Cl}(P))$ and whose other side is a hemisphere in $\mathbf{dev}(\mathrm{Cl}(\alpha_S)) \cup \mathbf{dev}(\mathrm{Cl}(\alpha_T))$. We can show easily that $\mathbf{dev}|S \cup T \cup \mathrm{Cl}(P)$ is an imbedding; hence, $S \cup T \cup \mathrm{Cl}(P)$ includes a radiant 3-bihedron U' mapping onto U. Since $\mathrm{Cl}(P) \subset U'$ and $\overline{vz}$ is transversal to F, it follows that $\overline{vz} - \{v\} \subset U'^o$, which means $y \in U'^o$; since $U' \sim R$ obviously, this contradicts $y \in \mathrm{bd}\Lambda(R)$. □

CHAPTER 6

2-convex radiant affine manifolds

We will now show that 2-convex radiant affine compact 3-manifolds with empty or totally geodesic boundary further decompose along totally geodesic affine tori or Klein bottles into convex radiant affine manifolds and concave-cone affine manifolds which are either Seifert spaces or finitely covered by a bundle over a circle with fiber homeomorphic to tori. (We assume that the Kuiper completion of the holonomy cover of the radiant affine 3-manifold does not include any 3-crescents or radiant bihedra.) As we said in the introduction, this major step will be accomplished in Chapters 6-10.

For this purpose, we will obtain a crescent-cone (Theorem 6.1) in this chapter, which will play the role of a crescent for real projective surfaces to give us our desired decomposition. First, we will define various objects in $\check{M}_h$. Next, we find a radiant tetrahedron F which detects the nonconvexity of M_h and name various parts to this tetrahedron.

Denoting by F_1, F_2, F_3, and F_4 the sides of the tetrahedron F, we find a sequence of points in the face F_3 of F which leaves every compact subset of M_h and equidistant from U and D, the upper and lower subsets of $F_1 \cup F_2$. We do the standard pull-back argument in [**11**]; we pull the points of the sequence by deck transformations to a fixed fundamental domain of M_h and pull F and the named subsets of F along with the points. Then by choosing subsequences, we assume that the each sequence of pulled-back images of the objects converges. We will show that the "limit" of the sequence of images of F is a nondegenerate 3-ball, i.e., a radiant trihedron or a radiant tetrahedron, and identify all the "limits" of the sequences of images of every object. (*Note that we will use the term "nondegenerate" merely to indicate it is not a lower dimensional object. The term is redundant actually but is used for emphasis.*)

Using the *claim*, i.e., Proposition 7.1, in the introduction to be proved in Chapters 7, 8, and 9, we verify at the end of the chapter:

THEOREM 6.1. *Let M be a compact 2-convex but nonconvex radiant affine 3-manifold with totally geodesic or empty boundary. Assume that the Kuiper completion $\check{M}_h$ of its holonomy cover M_h does not include radiant bihedra. Then $\check{M}_h$ includes a crescent-cone.*

We will use the following hypothesis of above theorem to be referred to in Chapters 6-10.

HYPOTHESIS 1. We assume that M is a 2-convex but non-convex radiant affine 3-manifold with totally geodesic boundary or empty boundary, and $\check{M}_h$ does not include 3-crescents or radiant bihedra.

We fix an identification of $\mathbf{R}^3$ with the interior of a closed 3-hemisphere $\mathcal{H}$, and denote by $\mathbf{S}^2_\infty$ the sphere at infinity, fix the development pair $\mathbf{dev} : \check{M}_h \to \mathbf{S}^3$ whose image lies in $\mathcal{H}$ as M is affine and $h : \pi_1(M)/\pi_1(M_h) \to \mathrm{Aut}(\mathcal{H})$ whose image fixes the origin O.

We recall some terminology: A *lune* in $\mathbf{S}^3$ is a closed convex subset of a great 2-sphere A in $\mathbf{S}^3$ that is the closure of a component of A with two distinct great circles in A removed; a *triangle* is the closure of a component of A with three great circles meeting in general position removed. A *trihedron* in $\mathbf{S}^3$ is a closed convex subset of $\mathbf{S}^3$ that is the closure of a component of $\mathbf{S}^3$ with three great 2-spheres meeting in general position removed. The boundary of a trihedron is the union of three lunes with disjoint interiors. A *tetrahedron* in $\mathbf{S}^3$ is a closed convex subset of $\mathbf{S}^3$ that is the closure of a component of $\mathbf{S}^3$ with four great 2-spheres meeting in general position removed. The boundary of a tetrahedron in $\mathbf{S}^3$ is the union of four triangles with disjoint interiors. Recall how the corresponding objects in $\check{M}_h$ and $\check{M}$ are defined.

REMARK 6.1. A convex set in $\mathrm{Cl}(\mathcal{H})$ is radiant if it is a cone over a convex set in $\mathbf{S}^2_\infty$. It is easy to see that given a hemisphere, a lune, or a simply convex triangle in $\mathbf{S}^2_\infty$, the cone over such sets are a radiant bihedron, a radiant trihedron, or a radiant tetrahedron respectively (see Remark 1.1).

DEFINITION 6.1. A *radiant finite lune* in $\check{M}_h$ is a lune A such that the boundary of $\mathbf{dev}(A)$ is the union of two edges, one of which passes through the origin and the other in $\mathbf{S}^2_\infty$. An *infinite lune* is a lune in $\check{M}$ included in $M^{\mathrm{i}}_{h\infty}$. A *crescent-cone* is a trihedron A in $\check{M}_h$ such the boundary of A is the union of three lunes, one of which is infinite, the second is a radiant finite lune and lies in $M_{h\infty}$ and the third is a radiant finite lune meeting M_h. (See Figure 6.1 and Example 6.1.)

EXAMPLE 6.1. Let x, y, z be the standard coordinates of $\mathbf{R}^3$. The holonomy cover of $\mathcal{E}_2$ for the three-dimensional case equals $U - l$ where U is the upper half-space given by $z \geq 0$ and l the z-axis. The closure of the subset of U given by $ax + by > 0$ for $a, b \in \mathbf{R}$ not both zero, is a crescent-cone. In fact the closure of the component missing l of U^o with a subspace of $\mathbf{R}^3$ meeting U^o removed is a crescent-cone in the Kuiper completion $\hat{\mathcal{E}}_2$ of the holonomy cover $U - l$ of $\mathcal{E}_2$. When N is a Benzécri suspension of a closed real projective surface Σ, the Kuiper completion of the holonomy cover N_h of N is homeomorphic to $(\check{\Sigma}_h \times I)/\sim$ for an interval $I = [0, 1]$ and the Kuiper completion $\check{\Sigma}_h$ of a holonomy cover Σ_h of Σ where the equivalent relation $\sim$ identifies $\Sigma \times \{0\}$ to a point. Here crescent-cones are precisely the closures of crescents times $(0, 1)$. Recall that crescents correspond to annuli in Σ. Hence, crescent-cones correspond to the Benzécri suspensions of the annuli.

We will now begin the pull-back process as in [**11**]. (To see examples, see Example 6.2). As a first step, we have the following:

PROPOSITION 6.1. *If M is not convex, then there exists a radiant tetrahedron F with two sides F_1, F_2 not in $M_{h\infty}$ and F_4 in $M^{\mathrm{i}}_{h\infty}$ and the remaining one F_3 such that F_3^o meets both M_h and $M_{h\infty}$ and $F_1 - (\{O\} \cup F_4)$ and $F_2 - (\{O\} \cup F_4)$ are subsets of M_h.*

PROOF. Let $\mathbf{S}^2_1$ denote the sphere of $\mathbf{d}$-radius $\pi/4$ in $\mathcal{H}$ with center O. Consider $C = \mathbf{dev}^{-1}(\mathbf{S}^2_1) \cap M_h$. Then C is connected since C meets every flow line in M_h. Since $\mathbf{S}^2_1$ has a flat real projective structure induced from the radial projection

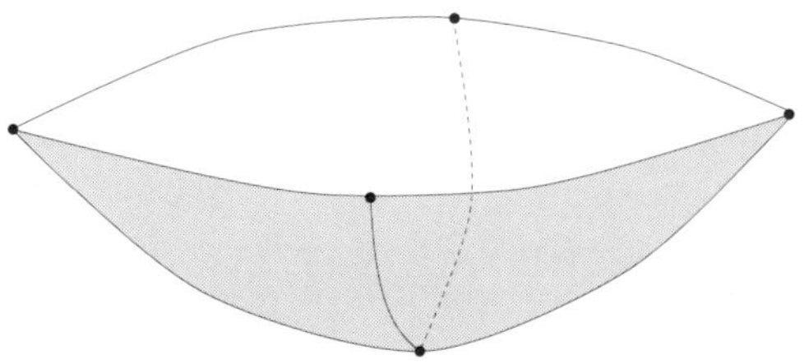

FIGURE 6.1. An example of a crescent-cone T, as a trihedron, where the grey area indicates the side in the closure of $T \cap M_{h\infty} - M^{\mathrm{i}}_{h\infty}$. The top side is in $M^{\mathrm{i}}_{h\infty}$. The picture is only schematic as the geodesics are bent.

$\mathbf{R}^3 - \{O\} \to \mathbf{S}^2_\infty$, the surface C has a real projective structure with geodesic boundary.

Suppose that C is not convex. $\mathbf{dev}|C : C \to \mathbf{S}^2_1$ can be considered a developing map of the real projective surface C. By inducing the metric from the Riemannian metric μ on $\mathbf{S}^3$ restricted on $\mathbf{S}^2_1$ to C and completing the induced path-metric on C (see [**15**]). Let $\check{C}$ denote the Kuiper completion of C and $\tilde{C}_\infty$ the set of ideal points. Since C is not convex, there exists a real projective triangle H in $\check{C}$ with an edge e of H satisfying $H \cap \tilde{C}_\infty = e^o \cap \tilde{C}_\infty \neq \emptyset$ by Theorem A.2 in [**15**].

Since the inclusion map $C \to M_h$ is a quasi-isometry, it follows that the $\check{C}$ may be regarded as a subset of $\check{M}_h$ so that $\tilde{C}_\infty$ is included in $M^{\mathrm{f}}_{h\infty}$.

Let J be the set $H \cap C$. The interior of the union J' of all radial lines in M_h through J is obviously a radiant open tetrahedron in M_h, and the closure F in $\check{M}_h$ is a radiant tetrahedron with a side $F_4 \subset M^{\mathrm{i}}_{h\infty}$. (To see this, apply $\Phi_{h,t}$ to an "ϵ-thickened" J for $t \in (-\infty, \infty)$ and take a union and apply Proposition 1.2.) Since two edges e_1 and e_2 of H other than e are in M_h, it follows that two radiant sides F_1 and F_2 of F with O and $M^{\mathrm{i}}_{h\infty}$ removed are subsets of M_h respectively. Since e is not a subset of M_h, it follows that the face F_3 corresponding to e meets M_h but is not a subset of M_h. Therefore, we found the desired radiant tetrahedron F.

If C is convex, then C is projectively diffeomorphic to a quotient of a convex domain in $\mathbf{S}^2_1$ by Proposition A.2 of [**15**]. As C admits a developing map $\mathbf{dev}|C$ to $\mathbf{S}^2_1$, it follows that C is projectively diffeomorphic to a convex domain in $\mathbf{S}^2_1$ by $\mathbf{dev}|C$. By a classification of convex sets in $\mathbf{S}^2_1$ (see [**15**]), $\mathbf{dev}(C)$ is either $\mathbf{S}^2_1$ itself, a hemisphere, a convex subset of a 2-hemisphere of $\mathbf{S}^2_1$. The final possibility implies that M_h is convex. Hence, C is projectively diffeomorphic to $\mathbf{S}^2_1$ or a hemisphere, and M_h to $\mathbf{R}^3 - \{O\}$ or a half-space with the origin removed by $\mathbf{dev}$. Thus $\check{M}_h$ can be identified with $\mathcal{H}$ or a closed half-space by $\mathbf{dev}$ where $M_{h\infty}$ is identifiable to $\{O\}$ union with $\mathbf{S}^2_\infty$ or a hemisphere in $\mathbf{S}^2_\infty$. We can easily find a tetrahedron in $\check{M}_h$ showing that M_h is not 2-convex (see Lemma 2.1 and Chapter 4 of [**15**]). □

We now produce notation which will be used extensively in this paper; they have fixed meanings in Chapters 6-9. Let v_i denote the vertex of F opposite F_i for each $i = 1, 2, 3, 4$, with $v_4 = O$; let I_{ij} for $i \neq j$ denote the edge of intersection of sides F_i and F_j. From above $F_3 \cap M_{h\infty}$ is the union of rays from the origin and the edge in $M^{\mathrm{i}}_{h\infty}$. Hence $F_3 - M_{h\infty}$ has two components G_1 and G_2 meeting with F_1 and F_2 at edges I_{13} and I_{23} respectively. $\mathrm{Cl}(G_1) \cap M_{h\infty}$ is the union of two segments

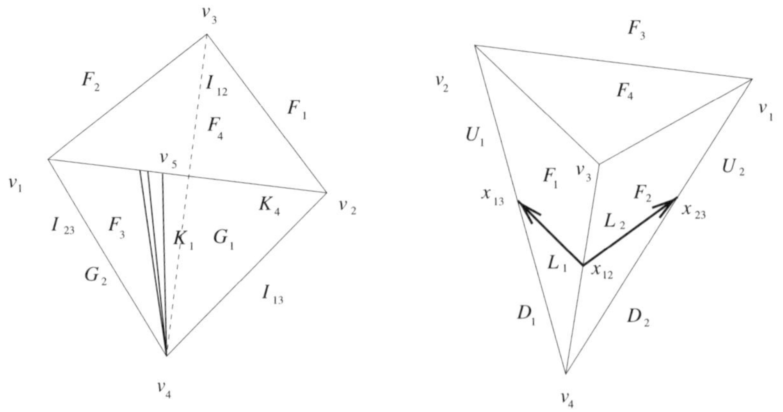

FIGURE 6.2. F, its sides, vertices, L_1, and L_2 viewed from two different points. (For convenience, we orient L_1 towards x_{13} and L_2 towards x_{23} and draw arrows to indicate directions.)

K_1 and K_4 where K_4 equals $\mathrm{Cl}(G_1) \cap F_4$ and K_1^o is a radial line in F_3^o. Let v_5 denote the intersection point of K_1 and K_4. The boundary of $\mathrm{Cl}(G_1)$ is the union of I_{12}, K_1 and K_4. Let us draw a segment L_1 on F_1 with endpoints x_{13} in I_{13}^o and x_{12} in I_{12}^o, and a segment L_2 in F_2 with endpoints x_{23} in I_{23}^o and x_{12}. We may assume that x_{ij} are midpoints of I_{ij} with respect to $\mathbf{d}$ for each pair i, j, $i = 1, \ldots, 4$ and $j = 1, \ldots, 4$. The closures of components of $F_1 - L_1$ are a quadrilateral and a triangle, which we denote by U_1 and D_1 respectively. Similarly we denote the closures of components of $F_2 - L_2$ by U_2 and D_2 respectively, and finally denote by U the set $U_1 \cup U_2$ and D the set $D_1 \cup D_2$. (See Figure 6.2.)

Recall the metric d_M on M_h induced from the path-metric d_M on M obtained from the Riemannian metric on M. For each point x of G_1, $d_M(x, D \cap M_h)$ is realized by a point $y \in D \cap M_h$ as $D \cap M_h$ is a properly imbedded in M_h; similarly, $d_M(x, U \cap M_h)$ is realized by a point $y \in U \cap M_h$. It follows from these facts that $d_M(\cdot, D \cap M_h)$ and $d_M(\cdot, U \cap M_h)$ are continuous functions on G_1. Let $\mathcal{E}$ be the set of points $x \in G_1$ satisfying $d_M(x, D \cap M_h) = d_M(x, U \cap M_h)$.

For a point x in the interior of the line $D \cap G_1 \cap M_h - L$, we obtain $d_M(x, D \cap M_h) < d_M(x, U \cap M_h)$ (see Figure 6.2). Similarly, for a point x in that of $U \cap G_1 \cap M_h - L$, we have $d_M(x, U \cap M_h) < d_M(x, D \cap M_h)$. Since $\mathcal{E}$ separates G_1 into two nonempty subsets where $d_M(x, D \cap M_h) > d_M(x, U \cap M_h)$ holds and where $d_M(x, D \cap M_h) < d_M(x, U \cap M_h)$ holds respectively, $\mathcal{E}$ is not a compact subset of G_1. Therefore, there exists a sequence of points ξ^i converging to a point ξ^∞ of $\mathrm{Cl}(G_1) \cap M_{h\infty} = K_1 \cup K_4$ (under the metric $\mathbf{d}$). Since $U \cap M_h$ and $D \cap M_h$ are properly imbedded submanifolds, we have a sequence $\{u^i\}$ in U and $\{d^i\}$ in D so that

$$d_M(\xi^i, u^i) = d_M(\xi^i, d^i).$$

Since u^i and d^i belong to the compact subset F of $\check{M}_h$, we may assume without loss of generality that $u^i \to u^\infty$ and $d^i \to d^\infty$ for points u^∞ and d^∞ in F. Choose the closure $\mathcal{W}$ of a fundamental domain of M_h and a deck transformation ϑ^i such that $\vartheta^i(\xi^i) = \eta^i$ for $\eta^i \in \mathcal{W}$.

We define

$$F^i = \vartheta^i(F), \quad F^i_j = \vartheta^i(F_j), \quad G^i_1 = \vartheta^i(\mathrm{Cl}(G_1)),$$
$$I^i_{jk} = \vartheta^i(I_{jk}), \quad L^i_l = \vartheta^i(L_l), \quad K^i_1 = \vartheta^i(K_1), \tag{6.1}$$
$$K^i_4 = \vartheta^i(K_4), \quad U^i_l = \vartheta^i(U_l), \quad \text{and } D^i_l = \vartheta^i(D_l)$$

for each $j, k = 1, 2, 3$ and $l = 1, 2$, and define x^i_{jk} to be $\vartheta^i(x_{jk})$ and v^i_l to be $\vartheta^i(v_l)$.

We will need the material on the Hausdorff convergence of compact subsets of $\mathbf{S}^n$ (see Appendix B).

HYPOTHESIS 2. By choosing subsequences, we may assume the following conditions. (Recall that $h(\vartheta_i) \circ \mathbf{dev} = \mathbf{dev} \circ \vartheta_i$.)

- η^i converges to a point η^u of $\mathrm{Cl}(\mathcal{W})$, i.e., $\eta^u \in M_h$.
- $$\mathbf{dev}(F^i) = h(\vartheta^i)(\mathbf{dev}(F)), \quad \mathbf{dev}(F^i_j) = h(\vartheta^i)(\mathbf{dev}(F_j)),$$
  $$\mathbf{dev}(I^i_{jk}) = h(\vartheta^i)(\mathbf{dev}(I_{jk})), \quad \mathbf{dev}(L^i_l) = h(\vartheta^i)(\mathbf{dev}(L_l)),$$
  $$\mathbf{dev}(G^i_1) = h(\vartheta^i)(\mathbf{dev}(\mathrm{Cl}(G_1))),$$
  $$\mathbf{dev}(K^i_1) = h(\vartheta^i)(\mathbf{dev}(K_1)), \quad \mathbf{dev}(K^i_4) = h(\vartheta^i)(\mathbf{dev}(K_4)),$$
  $$\mathbf{dev}(U^i_l) = h(\vartheta^i)(\mathbf{dev}(U_l)), \quad \text{and } \mathbf{dev}(D^i_l) = h(\vartheta^i)(\mathbf{dev}(D_l))$$

  for each $j, k = 1, 2, 3$ and $l = 1, 2$ converge geometrically to compact convex sets

  $$F^\infty, F^\infty_j, I^\infty_{jk}, L^\infty_l, G^\infty_1, K^\infty_1, K^\infty_4, U^\infty_l, \text{ and } D^\infty_l \tag{6.2}$$

  in $\mathbf{S}^3$ respectively.
- The sequence of outer-normal vectors n^i at η^i converges to a unit vector at η^u.
- $h(\vartheta^i)(\mathbf{dev}(x_{jk}))$ and $h(\vartheta^i)(\mathbf{dev}(v_l))$ converge to points x^∞_{jk} and v^∞_l respectively for each $j, k = 1, 2, 3$ and $l = 1, 2, 3, 4, 5$.

It follows easily that the dimension of F^∞ is ≤ 3, those of $F^\infty_j, G^\infty_1, U^\infty_l$, and $D^\infty_l \leq 2$, and those of I^∞_{jk}, K^∞_1, K^∞_4, L^∞_l and $U^\infty_l \leq 1$. These are all balls of such dimensions by Proposition 2.8 of [**15**]. (See the Appendix B).

For example, I^∞_{jk} is either a segment of $\mathbf{d}$-length $\leq \pi$ or a point since the limit I^∞_{jk} is convex. I^∞_{jk} for j, k not equal to 4 is a radial segment always. Since F^∞_l is a cone over I^∞_{l4} for $l = 1, 2, 3$ by Lemma B.1, it follows that F^∞_l is either a radial segment, a radiant triangle, or a radiant lune for $l = 1, 2, 3$.

EXAMPLE 6.2. We may do this process in $\mathcal{E}_2$ (see Example 1.1) in the three-dimensional case. Its holonomy cover equals $U - l$ where U is the upper half-space and l the z-axis. Then we can easily construct each of the objects in the process above. A total cross-section to the radial flow comes from in an affine plane parallel to the xy-plane. Hence components of the inverse image of the total cross-sections are affine planes parallel to the xy-plane. We may choose L_1 and L_2 to lie in a component J of such an inverse image. In this case, the set to be denoted by $\mathcal{E}$ may be considered also to be the intersection of G_1 with J. (Keep this example in mind for the rest of the process.)

Similar examples can be obtained from Benzécri suspensions of nonconvex real projective surfaces.

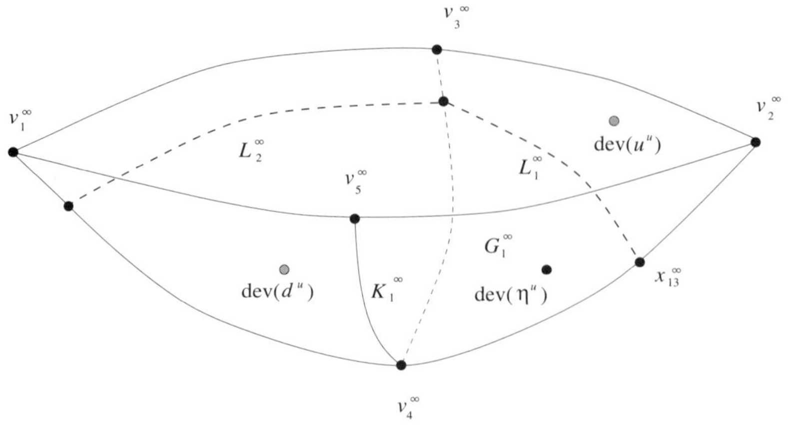

FIGURE 6.3. An example of a limit.

PROPOSITION 6.2. *F^i includes a common convex open ball $\mathcal{P}$ for sufficiently large i, and F^∞ is a 3-ball; that is, $h(\vartheta^i)(\mathbf{dev}(F))$ does not degenerate into a convex set of dimension less than or equal to 2.*

PROOF. Let $B(\eta^\infty)$ be a tiny ball of η^∞, and 2γ be the positive number equal to $d_M(\eta^\infty, \mathrm{bd}B(\eta^\infty))$. For i sufficiently large, $d_M(\eta^\infty, \eta^i) < \gamma$. We assume this holds for i in this proof.

Suppose we have either $d_M(\xi^i, u^i) < \gamma$ or $d_M(\xi^i, d^i) < \gamma$ for infinitely many i, either of of which means that $d_M(\eta^i, \vartheta^i(u^i)) < \gamma$ *and* $d_M(\eta^i, \vartheta^i(d^i)) < \gamma$ by equation 7.1. Since for each i, $\vartheta^i(u^i)$ and $\vartheta^i(d^i)$ belong to $B(\eta^\infty)$, there exists a segment s_i, $s_i \subset B(\eta^\infty)$ with these points as endpoints by convexity of $B(\eta^\infty)$; the segment $\vartheta^{i,-1}(s_i)$ is a segment in the compact set F with endpoints u_i and d_i. By choosing a subsequence if necessary, we may assume without loss of generality that the sequence of $\vartheta^{i,-1}(s_i)$ converges to a segment s with endpoints in U and D respectively. Such a segment s must have a point t belonging to $F_1 \cap M_h$ if endpoints are in U_1 and D_1 respectively; to $F_2 \cap M_h$ if endpoints in U_2 and D_2 respectively; and to $F^o \cup (F_3^o - M_{h\infty})$ otherwise by considering their images under **dev** and an elementary geometry argument (see Figure 6.4).

Since $B(\eta^\infty)$ is a compact subset of M_h, the images of $\vartheta^{i,-1}(s_i)$ may not intersect a compact subset of $F \cap M_h$ infinitely many times by Lemma 6.1. But a compact neighborhood of t in M_h intersects infinitely many of these, which is a contradiction. Therefore, for sufficiently large i, we have $d_M(\xi^i, u^i) \geq \gamma$ and $d_M(\xi^i, d^i) \geq \gamma$; that is, we have $d_M(\eta^i, \vartheta^i(F_1 \cup F_2)) \geq \gamma$. If i is sufficiently large so that $d_M(\eta^u, \eta^i) < \gamma/2$, then $\vartheta^i(F_1 \cup F_2 \cup F_4)$ does not meet the interior of the $\gamma/2$-ball A of η^u in $B(\eta^u)$.

Since η^i converges to η^u, and $\vartheta^i(F_1 \cup F_2 \cup F_4)$ does not intersect the interior of $\mathbf{dev}(A)$, Lemma B.2 implies the conclusion. □

LEMMA 6.1. *Let ϕ^i, $i = 1, 2, \ldots,$ be a sequence of distinct deck transformations of M_h, and K a compact subset of M_h. Then $\phi^i(K)$ intersects a compact ball neighborhood of a point of M_h for only finitely many i.*

PROOF. This follows since the action of the deck transformation group is properly discontinuous. □

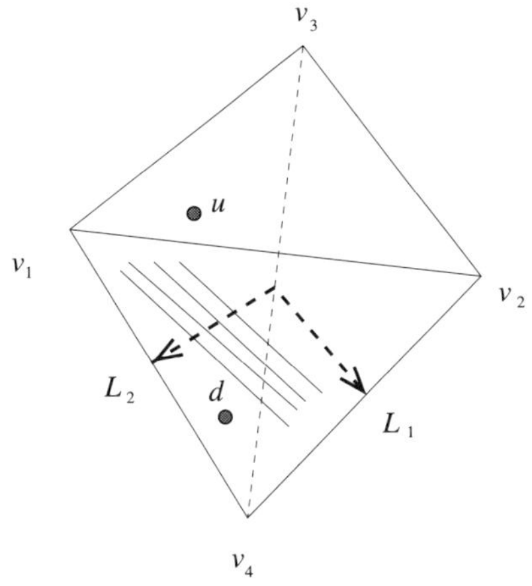

FIGURE 6.4. The segments indicate $\vartheta^{i,-1}(s_i)$.

Recall that a sequence of compact convex triangles in $\mathbf{S}^2$ converges to a point, a segment, a hemisphere, a lune, or a simply convex triangle (see Chapter 2 of [**15**] and the Appendix of [**11**].) Since F^∞ is a nondegenerate 3-ball by Proposition 6.2, and F^∞ is a cone over F_4^∞ by Lemma B.1 (see Definition 1.1 and Remark 1.1), F_4^∞ is either a hemisphere, a lune, or a simply convex triangle. By Remark 6.1, F^∞ is either a radiant bihedron, a radiant trihedron, or a radiant tetrahedron.

Similarly to the proof of Theorem 4.6 in [**15**], Theorem B.1 implies that there exists a convex 3-ball F^u in $\check{M}_h$ such that $\mathbf{dev}(F^u)$ equals F^∞ since we have a sequence of 3-balls F^i in $\check{M}_h$ including the common ball $\mathcal{P}$ by above Proposition 6.2.

Since $\mathbf{dev}|F^u$ is an imbedding onto F^∞, F^u includes subsets

$$F_i^u, I_{jk}^u, L_l^u, U_l^u, D_l^u$$

respectively corresponding to

$$F_i^\infty, I_{jk}^\infty, L_l^\infty, U_l^\infty, D_l^\infty$$

for each $i, j, k = 1, 2, 3, 4$, $j \neq k$, and $l = 1, 2$. Also, there are points x_{jk}^u corresponding to x_{jk}^∞ for each $j, k = 1, 2, 3$, $j \neq k$ and points v_j^u corresponding to v_j^∞ for $j = 1, 2, 3, 4, 5$. (See Figure 6.3.)

From the knowledge we have of objects with superscripts ∞, we gain informations about ones with superscripts u. For example, I_{jk}^u is a segment of $\mathbf{d}$-length $\leq \pi$ or a point, and F_i^u for $i = 1, 2, 3$ is a radial segment, a radiant triangle, or a radiant lune.

For the purposes of the later chapters, we state here:

REMARK 6.2. Notice that I_{12}^u, I_{13}^u, and I_{23}^u are radial segments. We have $I_{12}^u = I_{13}^u$ or $I_{12}^u \cap I_{13}^u = \{O\}$ respectively. Similarly, we have $I_{12}^u = I_{23}^u$ or $I_{12}^u \cap I_{23}^u = \{O\}$. The following facts are useful: L_1^u either equals O, is a segment with one endpoint in I_{12}^u and the another one in I_{13}^u, or a point in I_{12}^u where $I_{12}^u = I_{13}^u$. L_2^u equals O, is a segment with one endpoint in I_{12}^u and the another one in I_{23}^u, or a point in I_{12}^u where $I_{12}^u = I_{23}^u$. The endpoints of L_1^u and L_2^u in I_{12}^u are identical. These follow from looking at L_1^∞, L_2^∞, and so on. (Here, the endpoint of a point is itself.)

If F^∞ is a radiant bihedron, then F^u has a side in $M_{h\infty}$, and since F^u is not equal to $\check{M}_h$, it follows that a side of F^u intersects M_h and F^u must be a 3-crescent. Since 3-crescents do not exist in $\check{M}_h$ by Hypothesis 1, we obtain:

PROPOSITION 6.3. *F^u is either a radiant trihedron or a radiant tetrahedron. F^u includes $\mathcal{P}$.* □

THE PROOF OF THEOREM 6.1. Since $(F_1 \cup F_2) \cap M_h$ equals $(U \cup D) \cap M_h$, by Proposition 7.1, i.e. the claim in the introduction, we can choose a sequence of points ξ_i in $\mathcal{E}$ so that $d_M(\xi_i, (F_1 \cup F_2) \cap M_h)$ converges to $+\infty$. Using the pull-back argument as in Chapter 6, we assume Hypothesis 2. By Theorem B.1, we obtain a radiant tetrahedron F^u in $\check{M}_h$ with sides F^u_1, F^u_2, F^u_3, and F^u_4 or a radiant trihedron F^u with boundary $F^u_1 \cup F^u_2 \cup F^u_3 \cup F^u_4$. But as F^i_1 and F^i_2 are ideal sequences by Proposition 7.1, it follows from Theorem B.1 that $F^u_1 \cup F^u_2$ is a subset of $M_{h\infty}$.

DEFINITION 6.2. A radiant tetrahedron T in $\check{M}_h$ which has three sides in $M_{h\infty}$ is said to be a *pseudo-crescent-cone.*

By Lemma 3.6, there is a side of F^u meeting M_h. Since F^i_4 is a subset of $M^i_{h\infty}$, $F^u_1 \cup F^u_2 \cup F^u_4$ is a subset of $M_{h\infty}$; F^u_3 is the only totally geodesic disk in δF intersecting with M_h. Thus if F^u is a radiant trihedron, then F^u is a crescent-cone (see Definition 6.1). If F^u is a radiant tetrahedron, then F^u is a pseudo-crescent-cone by Definition 6.2.

In Chapter 12, we will show that $\check{M}_h$ cannot have a pseudo-crescent-cone since M is not convex. □

CHAPTER 7

The claim and the rooms

We will be proving the *claim* in the introduction, which will be proved in this chapter and Chapters 8 and 9 completely: We assume otherwise and find contradictions. The purpose of this chapter is to list properties which will be used in Chapters 8 and 9: Propositions 7.2, 7.3,7.4, and 7.5, and Remark 7.1.

PROPOSITION 7.1. *The supremum of* $d_M(x, D \cap M_h)$ *on* $\mathcal{E}$ *is infinite.*

We will prove this by contradiction: we assume the following hypothesis from now on (in this chapter and Chapters 8 and 9):

HYPOTHESIS 3. Let C_{ud} denote the finite supremum of $d_M(x, D \cap M_h)$ for $x \in \mathcal{E}$.

To show contradiction, let ξ^i be a sequence of points converging to a point ξ^∞ of $\mathrm{Cl}(G_1) \cap M_{h\infty} = K_1 \cup K_4$ (under the metric $\mathbf{d}$) as in the above chapter satisfying the Hypothesis 2. $d_M(\xi^i, D \cap M_h)$ is bounded above; we have a sequence $\{u^i\}$ in U and $\{d^i\}$ in D so that

$$d_M(\xi^i, u^i), d_M(\xi^i, d^i) < C_{ud} + 1 \text{ for all } i, \text{ and } d_M(\xi^i, u^i) = d_M(\xi^i, d^i). \tag{7.1}$$

We will use the notations of previous chapters but we will have some additional requirements: Let $\mathcal{W}'$ the $C_{ud}+1$ d_M-neighborhood of $\mathcal{W}$, to which $\vartheta^i(u^i)$ and $\vartheta^i(d^i)$ belong. In addition to Hypothesis 2, we will also require:

- $\vartheta^i(u^i)$ and $\vartheta^i(d^i)$ converge to points u^u and d^u in $\mathcal{W}'$ respectively.

From the previous chapter, $\mathbf{dev}(F^i)$ converges to F^∞, F^i shares a common open ball $\mathcal{P}$, and $\check{M}_h$ includes a radiant tetrahedron or radiant trihedron F^u including $\mathcal{P}$ so that $\mathbf{dev}(F^u) = F^\infty$. All associated objects are the same as in the chapter.

Since M has empty or totally geodesic boundary, each point x of M_h has an open coordinate chart U such that a lift $\phi : U \to \mathbf{S}^n$ of a chart $U \to \mathbf{R}P^n$ so that $\phi(U)$ is a convex set in $\mathcal{H}^o$. Hence, U includes a compact convex n-ball neighborhood B of x such that $\phi|B$ is an imbedding onto $\phi(B)$. Since $\mathbf{dev}$ is a continuation of charts, $\mathbf{dev}|B$ is a restriction of a chart, and hence, $\mathbf{dev}|B$ is an imbedding onto a convex compact n-ball. We say that B a *tiny ball neighborhood* of x. We can easily show that B can be chosen so that $\mathbf{dev}(B)$ is a $\mathbf{d}$-ball of certain radius, perhaps intersected with a closed affine half-space in $\mathcal{H}^o$ if B meets the totally geodesic boundary δM_h.

Next, we give some information on η^u, d^u, and u^u. Recall that every point x of M_h has a tiny-ball neighborhood $B(x)$. Since $\mathcal{W}'$ is compact and M_h has totally geodesic boundary, using Lebesgue number, we may assume without loss of generality that each point x of $\mathcal{W}'$ has a tiny-ball neighborhood $B(x)$ such that the image $\mathbf{dev}(B(x))$ is a $\mathbf{d}$-ball of constant radius ϵ, $\epsilon > 0$, perhaps intersected with a closed half-space of $\mathcal{H}^o$.

LEMMA 7.1. *Points η^u, u^u, and d^u have tiny-ball neighborhoods $B(\eta^u)$, $B(u^u)$ and $B(d^u)$. $\mathbf{dev}(B(\eta^u))$, $\mathbf{dev}(B(u^u))$, and $\mathbf{dev}(B(d^u))$ are uniformly bounded away from O and $\mathbf{S}^2_\infty$.*

PROOF. $\mathbf{dev}(M_h)$ is disjoint from O by Lemma 3.1. Since $\mathbf{dev}(M_h)$ is a subset of $\mathcal{H} - (\mathbf{S}^2_\infty \cup \{O\})$, $\mathbf{dev}(\mathcal{W}')$ is a compact subset of $\mathcal{H} - (\mathbf{S}^2_\infty \cup \{O\})$, and the last statement follows. □

By an upper component of $F^u_l - L^u_l$, we mean one further away from O than L^u_l; and by a lower component, we mean the one closer to O than L^u_l.

PROPOSITION 7.2.
- *Suppose that F^u_l is a radiant triangle or a radiant lune. U^u_l equals*
  - *either the union of the upper component of $F^u_l - L^u_l$ and L^u_l, or equals L^u_l itself when $F^u_l - L^u_l$ has only the lower component; and*
  - *D^u_l equals either the union of the lower component of $F^u_l - L^u_l$ and L^u_l, or equals L^u_l itself when $F^u_l - L^u_l$ has only the upper component.*
- *If F^u_l is a radial segment, then*
  - *either U_l is the closure of $F^u_l - \{p\}$ for a lower endpoint p of L^u_l if p does not equal the upper endpoint p of F^u_l, or U_l equals the upper endpoint of F^u_l otherwise ; and*
  - *either D_l is the closure of $F^u_l - \{p\}$ for an upper endpoint p of L^u_l if p does not equal O, or D_l equals $\{O\}$ otherwise.*

PROOF. Since U^i_1 is the closure of the upper component of $F^i_l - L^i_l$, the geometric limit U^∞_l is as stated above. □

LEMMA 7.2. *We have $\eta^u \in F^u$, and $d^u, u^u \in F^u$. Moreover, $\eta^u \in G^u_1$, $u^u \in F^u_l$, and $d^u \in F^u_{l'}$ for some l, $l' = 1, 2$.*

PROOF. We know that $u^i \in F^i$ for each i. Since $u^i \to u^u$, it follows that $u^i \subset \mathrm{int} B(u^u)$ whenever i is sufficiently large for a tiny ball $B(u^u)$ of u^u. Hence, F^i and $B(u^u)$ overlap and $\mathbf{dev}|F^i \cup B(u^u)$ is an imbedding onto $\mathbf{dev}(F^i) \cup \mathbf{dev}(B(u^u))$ by Proposition 1.2. Since F^i and F^u both include $\mathcal{P}$ for i sufficiently large, $\mathbf{dev}|F^i \cup F^u$ is an imbedding onto $\mathbf{dev}(F^i) \cup \mathbf{dev}(F^u)$. By choosing $B(u^u)$ carefully, we may assume that $(\mathbf{dev}(F^i) \cup \mathbf{dev}(B(u^u))) \cap (\mathbf{dev}(F^i) \cup \mathbf{dev}(F^u))$ is a connected submanifold with interior equal to $(\mathbf{dev}(F^i) \cup \mathbf{dev}(B(u^u)))^o \cap (\mathbf{dev}(F^i) \cup \mathbf{dev}(F^u))^o$, which can be accomplished by making these sets star-shaped from a point of $\mathbf{dev}(F^i)$. Since Proposition 1.2 shows that $\mathbf{dev}|F^i \cup F^u \cup B(u^u)$ is an imbedding, and $\mathbf{dev}(u^i) \to \mathbf{dev}(u^u)$, and $\mathbf{dev}(u^u) \in \mathbf{dev}(F^u)$, we conclude that $u^u \in F^u$. Similarly, we obtain $\eta^u \in F^u$ and $d^u \in F^u$.

Since $\mathbf{dev}(u^i) \in \mathbf{dev}(F^i_m)$ for each i and some m, $m = 1, 2$, depending on i, we have $\mathbf{dev}(u^u) \in \mathbf{dev}(F^u_l)$ for some l, $l = 1, 2$. Since $u^u \in F^u$, we have $u^u \in F^u_l$. Similarly, we obtain $\eta^u \in G^u_1$, and $d^u \in F^u_{l'}$ for some l', $l' = 1, 2$. □

LEMMA 7.3. *Suppose that a point x of M_h belongs to F^u. Then $\mathbf{dev}|F^u \cup B(x)$ is an imbedding for a choice of the tiny ball $B(x)$ of x, and $\mathbf{dev}|F^u \cup B(x) \cup F^i$ is an imbedding onto $F^\infty \cup B(x) \cup \mathbf{dev}(F^i)$ for i sufficiently large.*

PROOF. We choose $B(x)$ so that $\mathbf{dev}(B(x)) \cup F^\infty$ is a star-shaped set from a point of $\mathbf{dev}(\mathcal{P})$ which is the common open ball in F^i from the proof of Proposition 6.2. The lemma now follows from Proposition 1.2. □

PROPOSITION 7.3. *The set $L_1^u \cup L_2^u$ is a subset of $M_{h\infty}$. The set $F_l^u \cap B(u^u)$ is a subset of U_l^u with nonempty relative interior in F_l^u if $u^u \in F_l^u$ and $F_l^u \cap B(d^u)$ is that of D_l^u with nonempty relative interior in F_l^u if $d^u \in F_l^u$. In particular, U_l^u includes a relatively open subset of F_l^u if $u^u \in F_l^u$, and D_l^u includes a relatively open subset of F_l^u if $d^u \in F_l^u$.*

PROOF. The set $L_1^i \cup L_2^i$ equals $\vartheta^i(L_1 \cup L_2)$, and $L_1 \cup L_2$ is a compact subset of M_h. If $L_1^u \cup L_2^u$ contains a point x of M_h, then x belongs to F^u. Choose the tiny ball $B(x)$ as in Lemma 7.3 so that $\mathbf{dev}|F^u \cup B(x) \cup F^i$ is a homeomorphism onto their images. Since a sequence of points p_i, $p_i \in \mathbf{dev}(L_1^i \cup L_2^i)$, converges to $\mathbf{dev}(x)$ in $\mathbf{S}^3$, Lemma 7.3 shows that $\mathbf{dev}(L_1^i) \cup \mathbf{dev}(L_2^i)$ intersects $\mathbf{dev}(B(x))$ for sufficiently large i, and hence so does $L_1^i \cup L_2^i$ with $B(x)$. This contradicts Lemma 6.1, and the first statement follows.

Since $B(u^u)$ does not meet $L_1^u \cup L_2^u$, so does not $B(u^u) \cap F_l^u$. Since u^u belongs to U_l^u, $B(u^u) \cap F_l^u$ is included in U_l^u by the connectedness of U_l^u implied by Proposition 7.2. The same argument works for $B(d^u)$. The last statement follows easily from this. □

PROPOSITION 7.4. *If $u^u \in F_1^u$, then $d^u \in F_2^u$, and if $u^u \in F_2^u$, then $d^u \in F_1^u$. Moreover, no L_l^u passes through $F_l^{u,o}$ for each l, $l = 1, 2$. Finally, F_l^u is a (nondegenerate) radiant triangle or lune for l, $l = 1, 2$.*

PROOF. Recall that F_l^u is either a radial segment, a radiant triangle, or a radiant lune for $l = 1, 2$. Suppose that F_l^u is a radiant triangle or lune and u^u and d^u belong to F_l^u for a given l. Then both U_l^u and D_l^u are convex 2-balls with nonempty relative interiors and are the closures of the components of $F_l^u - L_l^u$. In order for this to happen, L_l^u passes through $F_l^{u,o}$ with endpoints in the radial sides of F_l^u (see Remark 6.2). Since this implies that every segment from O to a point of I_{l4}^u intersects L_l^u, $F_l^{u,o}$ is a radiant set, and $F_l^{u,o} \cap M_{h\infty}$ is a radiant set, it follows that $F_l^u \subset M_{h\infty}$. This contradicts the fact that U_l^u contains a point u^u, a point of M_h. (See Figure 6.3.)

We can show similarly that the situation where F_l^u is a radial segment and $u^u, d^u \in F_l^u$ does not occur also. This implies the first part of the lemma.

If L_l^u passes through $F_l^{u,o}$ for some l, then F_l^u is a subset of $M_{h\infty}$ as shown in the above paragraph. Hence, F_k^u for $k \neq l$, $k = 1, 2$, must contain both u^u and d^u, points of M_h. This is a contradiction by the first statement of this lemma, and the second part of the lemma is proved.

Suppose that F_1^u is a radial segment. Since F^u is a "nondegenerate" ball, F_2^u has to be a radiant lune and F_1^u is an edge of F_2^u. By above paragraph, F_2^u cannot contain both u^u and d^u. But if u^u belongs to F_1^u and d^u to F_2^u, then u^u also belongs to F_2^u, which is a contradiction. Similarly a contradiction follows if d^u belongs to F_1^u and u^u to F_2^u.

Analogously, F_2^u cannot be a radial segment also. □

REMARK 7.1. By above, F_l^u are disks always for $l = 1, 2$. If $u^u \in F_l^u$, then U_l^u includes an open subset of F_l^u by Proposition 7.3, and U_l^u is a nondegenerate disk. If $d^u \in F_l^u$, then D_l^u is a nondegenerate disk. In particular, U^u and D^u must have nonempty interior in the manifold-boundary δF^u of the 3-ball F^u.

PROPOSITION 7.5. *There exists a compact nondegenerate 2-disk G_1^u in F^u corresponding to G_1^∞, F_3^u is a nondegenerate 2-disk, δG_1^u includes segments K_1^u and K_4^u corresponding to K_1^∞ and K_2^∞. Moreover K_1^u and K_2^u are subsets of $M_{h\infty}$.*

PROOF. A sequence η^i converges to η^u, which has a tiny ball $B(\eta^u)$. We assume that $\eta^i \in \mathrm{int}B(\eta^u)$ for each i, and $\mathbf{dev}|F^i \cup B(\eta^u)$ is an imbedding onto its image by Proposition 1.1. Let J^i be the maximal totally geodesic subsurface in M_h including the totally geodesic subsurface $G_1^i \cap M_h$ so that $J^i \cap B(\eta^u)$ is a convex ball D^i with boundary in $\delta B(\eta^u)$.

Since K_1^i and K_4^i are subsets of $M_{h\infty}$, they don't intersect $B(\eta^u)$, and only the interior of the edge I_{13}^i of G_1^i may meet with $B(\eta^u)$. Let l^i be the segment $I_{13}^i \cap B(\eta^u)$ which has endpoints in $D^i \cap \mathrm{bd}B(\eta^u)$ if $I_{13}^i \cap B(\eta^u)$ is not empty; otherwise, let $l^i = \emptyset$. Since no other segment of G_1^i meets with $B(\eta^u)$, and η^i belongs to G_1^i, it follows that $G_1^i \cap B(\eta^u)$ is the closure of a component D'^i of the convex disk D^i with l^i removed, and the closure of D'^i is a compact convex disk bounded by l^i and an arc α^i in $\mathrm{bd}B(\eta^u)$.

Since $\mathbf{d}(\eta^u, \mathrm{bd}B(\eta^u))$ is bounded from below by a positive constant c, we obtain $\mathbf{d}(\eta^u, \alpha^i) \geq c$. Since $\eta^i \to \eta^u$, we obtain $\mathbf{d}(\eta^i, \alpha^i) > c/2$ for i sufficiently large. Since D'^i contains η^i for each i, an elementary geometry shows that D'^i includes a $\mathbf{d}$-disk of radius $c/2$ for each i (using the same idea as in the proof of Lemma B.2) implying that $\mathbf{dev}(G_1^i)$ is not degenerating into a segment, the first statement. The rest follows from Theorem B.1. □

CHAPTER 8

The radiant tetrahedron case

The aim of Chapters 8 and 9 is to show that Hypothesis 3 leads to contradictions under the following assumptions:

- F^u is a radiant tetrahedron.
- F^u is a radiant trihedron.

Since these two are the only possible shapes of F^u by Proposition 6.3, we will have completed the proof of Proposition 7.1.

We will show that the first case does not happen in this chapter by singling out two cases by considering L_l^u and U_l^u and D_l^u using mainly Propositions 7.3 and 7.4. Then we show that the two cases cannot happen by showing that their existence implies the existence of radiant bihedra, which was ruled out by our hypothesis 1. We obtain the radiant bihedra by putting our tetrahedra in standard positions and estimating the eigenvalues of holonomy action asymptotically from the positions of L_1^u and L_2^u. (The proof in this case will be a prototype of all later cases which will occur.)

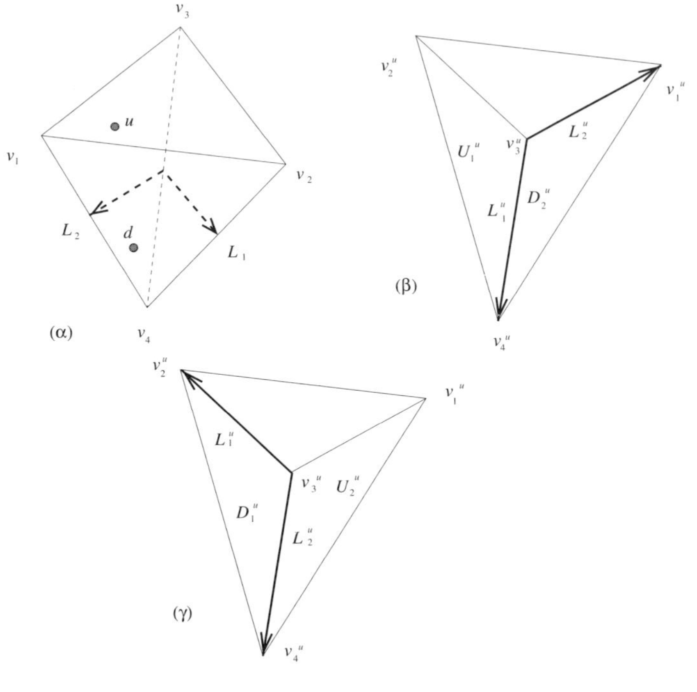

FIGURE 8.1. The objects of F, and two examples of F^u

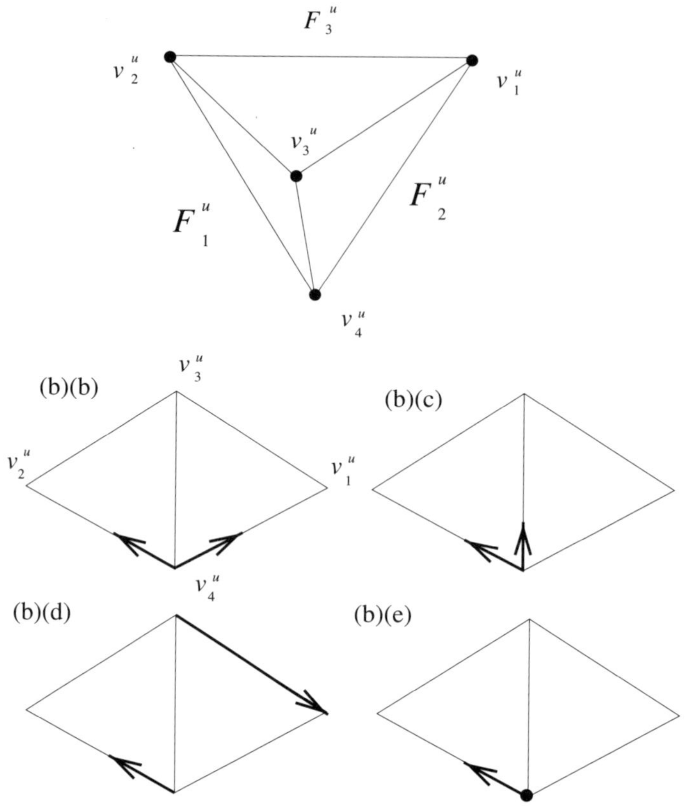

FIGURE 8.2. The arrows indicate the orientations.

REMARK 8.1. Whenever points x and y both belong to F^i (resp. F^u) for a fixed i and $\mathbf{dev}(x)$ and $\mathbf{dev}(y)$ not antipodal, we denote by $\overline{xy}$ the unique segment in F^i (resp. F^u) connecting x and y.

To begin, F^∞ is a triangle cone with vertices v_i^u for $i = 1, 2, 3, 4$, each F_l^u is a nondegenerate triangle which forms sides of F^∞, and each I_{jk}^u is a segment forming the sides of F_l^u for some l. Since L_1^u is a segment or a point with one endpoint in I_{12}^u and the another one in I_{13}^u (see Remark 6.2), L_1^u satisfies one of the following mutually exclusive statements:

(a) L_1^u passes a point of $F_1^{u,o}$.
(b) L_1^u equals a segment in I_{13}^u with an endpoint O.
(c) L_1^u equals a segment in I_{12}^u with an endpoint O.
(d) L_1^u equals I_{14}^u.
(e) L_1^u equals the point O.

Similarly, L_2^u satisfies one of the following mutually exclusive statement:

(a) L_2^u passes a point of $F_2^{u,o}$.
(b) L_2^u equals a segment in I_{23}^u with an endpoint O.
(c) L_2^u equals a segment in I_{12}^u with an endpoint O.
(d) L_2^u equals I_{24}^u.
(e) L_2^u equals the point O.

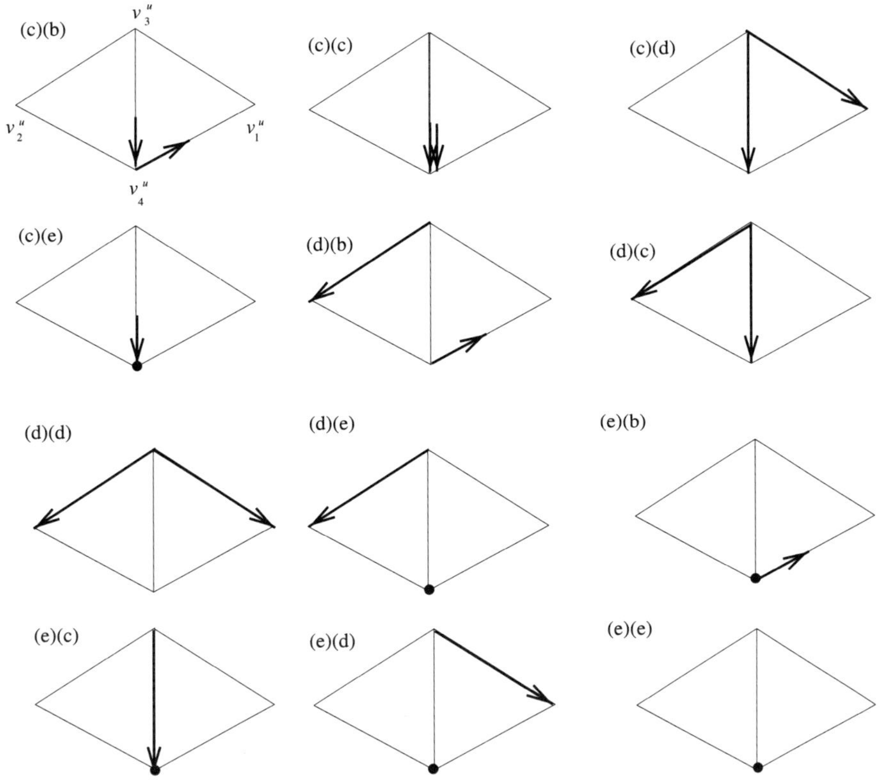

FIGURE 8.3.

We list out all possibilities of L_1^u and L_2^u in Figures 8.2 and 8.3. By Proposition 7.4, L_l^u can be as in (b)-(e) and L_1^u and L_2^u share an endpoint x_{12}^u. By these two conditions and Remark 7.1, we can have only one of the following possibilities:

(i) $F_1^u = U^u$, $F_2^u = D^u$, $L_1^u = \overline{v_3^u O}$, and $L_2^u = \overline{v_1^u v_3^u}$ ((c)(d) in Figure 8.3), or
(ii) $F_1^u = D^u$, $F_2^u = U^u$, $L_1^u = \overline{v_2^u v_3^u}$, and $L_2^u = \overline{v_3^u O}$ ((d)(c) in Figure 8.3).

DEFINITION 8.1. Given two sequences of positive numbers a_i and b_i with

$$\lim_{i \to \infty} a_i/b_i \in [0, \infty],$$

we say $a_i \gg b_i$ if $a_i/b_i \to \infty$. We say $a_i \sim b_i$ if $|\log(a_i/b_i)|$ is bounded.

Given any pair of sequences of positive numbers, up to a choice of subsequences, one of the three must hold.

LEMMA 8.1. *Let s be a segment in $\mathbf{S}^n$ of $\mathbf{d}$-length $< \pi$, and p a point of s^o. Let $\{\vartheta_i\}$ be a sequence of real projective transformations acting on s fixing each endpoints x and y of s. Then $\vartheta_i(p)$ converges to x if and only if $\lambda_i \gg \nu_i$ for the eigenvalue λ_i and ν_i corresponding to x and y of ϑ_i. Moreover, if $\vartheta_i(p)$ is convergent, then the limit is in s^o if and only if $\lambda_i \sim \nu_i$.*

PROOF. The proof is obvious. □

TABLE 1. The lower table shows how to obtain various objects from another objects in the upper table.

(1)	(2)	(3)	(4)	(5)
$\mathbf{dev}(F)$	$\mathbf{dev}(F^i)$	$\psi(\mathbf{dev}(F))$	$\psi^i(\mathbf{dev}(F^i))$	$\psi^\infty(\mathbf{dev}(F^u))$
$\mathbf{dev}(F_j)$	$\mathbf{dev}(F^i_j)$	$\psi(\mathbf{dev}(F_j))$	$\psi^i(\mathbf{dev}(F^i_j))$	$\psi^\infty(\mathbf{dev}(F^u_j))$
$\mathbf{dev}(L_l)$	$\mathbf{dev}(L^i_l)$	$\psi(\mathbf{dev}(L_l))$	$\psi^i(\mathbf{dev}(L^i_l))$	$\psi^\infty(\mathbf{dev}(L^u_l))$
$\mathbf{dev}(x_{jk})$	$\mathbf{dev}(x^i_{jk})$	$\psi(\mathbf{dev}(x_{jk}))$	$\psi^i(\mathbf{dev}(x^i_{jk}))$	$\psi^\infty(\mathbf{dev}(x^u_{jk}))$
$\mathbf{dev}(\mathrm{Cl}(G_1))$	$\mathbf{dev}(G^i_1)$	$\psi(\mathbf{dev}(\mathrm{Cl}(G_1)))$	$\psi^i(\mathbf{dev}(G^i_1))$	$\psi^\infty(\mathbf{dev}(G^u_1))$
$\mathbf{dev}(K_1)$	$\mathbf{dev}(K^i_1)$	$\psi(\mathbf{dev}(K_1))$	$\psi^i(\mathbf{dev}(K^i_1))$	$\psi^\infty(\mathbf{dev}(K^u_1))$

$h(\vartheta^i)$	ψ	$\psi^i_\&$	ψ^i	convergence
(1) $\longrightarrow$ (2)	(1) $\longrightarrow$ (3)	(3) $\longrightarrow$ (4)	(2) $\longrightarrow$ (4)	(4) $\longrightarrow$ (5)

There exist projective automorphisms ψ and ψ^i such that $\psi(\mathbf{dev}(v_j))$ and $\psi^i(\mathbf{dev}(v^i_j))$ for $j = 1, 2, 3, 4$ are in standard positions, i.e., at

$$[1, 0, 0, 0], [0, 1, 0, 0], [0, 0, 1, 0], \text{ and } [0, 0, 0, 1]$$

the vertices of the standard tetrahedron T_s in $\mathbf{S}^3$, respectively, such that ψ^i form a bounded sequence in $\mathrm{Aut}(\mathbf{S}^3)$, which can be obtained since $\mathbf{dev}(v^i_j)$ converges to the vertices of F^∞ a nondegenerate tetrahedron. By choosing subsequences, we may assume that $\psi^i \to \psi^\infty$ for some real projective automorphism ψ^∞ in $\mathrm{Aut}(\mathbf{S}^3)$. Since $\mathbf{dev}(F^i) \to F^\infty$, it follows that $\psi^\infty(\mathbf{dev}(F^u))$ is in a standard position, i.e., $\psi^\infty(\mathbf{dev}(F^u)) = T_s$.

We also note that

$$\psi^i_\& = \psi^i \circ h(\vartheta^i) \circ \psi^{-1} \tag{8.1}$$

fixes the vertices of the standard tetrahedron T_s. Such a projective automorphism has a diagonal matrix expression with entries nonnegative; let $\lambda^i_1, \dots, \lambda^i_4$ be the eigenvalues of $\psi^i_\&$ corresponding to

$$[1, 0, 0, 0], [0, 1, 0, 0], [0, 0, 1, 0], \text{ and } [0, 0, 0, 1]$$

respectively. For convenience, we assume that $\psi(x_{12})$, $\psi(x_{23})$, and $\psi(x_{13})$ are at $[0, 0, 1, 1]$, $[1, 0, 0, 1]$, and $[0, 1, 0, 1]$ respectively. In table 1, we summarized the relation between all objects, which we will continue to use for situations in Chapters 8, and 9.

We will show (i) cannot happen. Suppose (i) happened. Then L^u_1 equals I^u_{12} and L^u_2 equals I^u_{24}. Hence, $x^u_{13} = O$, $x^u_{12} = v^u_3$, and $x^u_{23} = v^u_1$. (See Figure 8.4.) Lemma 8.1 and Table 1 imply that

$$\lambda^i_4 \gg \lambda^i_2, \lambda^i_1 \gg \lambda^i_4, \lambda^i_3 \gg \lambda^i_4 \tag{8.2}$$

since

$$x^\infty_{13} = O, x^\infty_{23} = v^\infty_1, \text{ and } x^\infty_{12} = v^\infty_3$$

respectively. Hence, λ^i_2 form the least eigenvalue sequence of $\psi^i_\&$.

Choose a small neighborhood $B(x)$ of a point $x \in I^o_{13}$. Then since $B(x) \in M_h$, every ray through $B(x)$ belongs to M_h; let C_x be the cone formed from the union of these rays. Since $\psi \circ \mathbf{dev}(\mathrm{Cl}(C_x))$ includes an open cone including $\overline{[0, 0, 0, 1][0, 1, 0, 0]}$, the condition on the eigenvalues implies that $\psi^i_\&(\psi \circ \mathbf{dev}(\mathrm{Cl}(C_x)))$ converges to a radiant bihedron including $\psi^\infty(F^\infty) = T_s$. As $\psi^i_\&(\psi \circ \mathbf{dev}(\mathrm{Cl}(C_x)))$

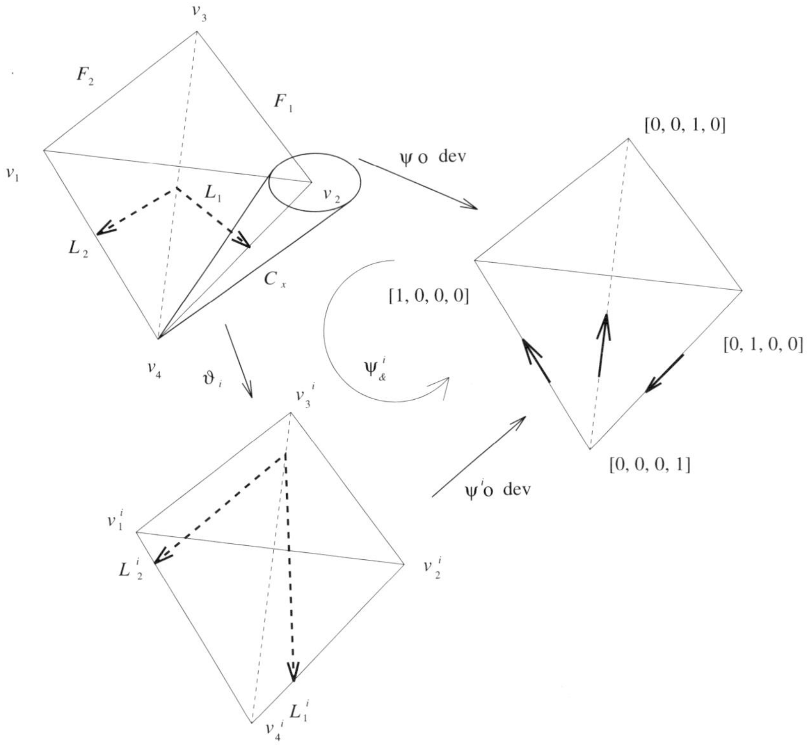

FIGURE 8.4. Case (i).

equals $\psi^i \circ \mathbf{dev}(\vartheta^i(\mathrm{Cl}(C_x)))$, and ψ^i is a convergent sequence, $\mathbf{dev}(\vartheta^i(\mathrm{Cl}(C_x))))$ converges to a radiant bihedron C^∞, which includes F^∞.

Since $\vartheta^i(\mathrm{Cl}(C_x))$ overlaps with $\vartheta^i(F)$ for each i, by the dominating part of Theorem B.1, there exists a radiant bihedron including F^u. This contradicts our hypothesis 1, and the case (i) cannot happen.

We will now show that (ii) does not happen. Suppose that (ii) happened. Then Lemma 8.1 and Table 1 imply that $\lambda_2^i \gg \lambda_4^i$ since $x_{13}^\infty = v_2^u$; $\lambda_4^i \gg \lambda_1^i$ since $x_{23}^\infty = v_4^u$; and $\lambda_3^i \gg \lambda_4^i$ since $x_{12}^\infty = v_3^\infty$. Hence, λ_1^i is the least eigenvalue sequence of $\psi^i_\&$, we obtain $v_5^i \to v_2^u$ as $\lambda_2^i \gg \lambda_1^i$, and, in particular, $h(\vartheta_i)\mathbf{dev}(\mathrm{Cl}(G_1))$ converges to a segment, which is a contradiction by Proposition 7.5. (See Figure 8.5.)

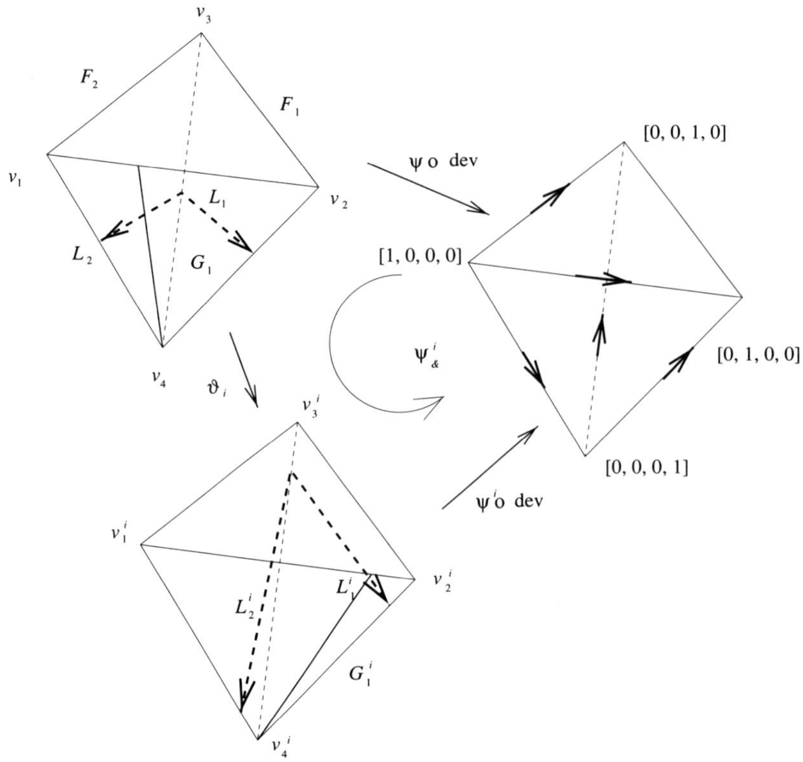

FIGURE 8.5. Case (ii).

CHAPTER 9

The radiant trihedron case

When F^u is a radiant trihedron, we divide our cases to (A), (B), and (C) depending on the locations of the vertices v_1^u, v_2^u, and v_3^u (see below). These cases will be ruled out in a manner similar to the above section for triangle cones; i.e., we use information on L_l^u and U^u and D^u to restrict our configurations, and using an asymptotic eigenvalue estimation for holonomy actions, we can either obtain a radiant bihedron, which is a contradiction as before, or obtain other contradictions straight away.

Assume that F^u is a trihedron with a face lune in $M_{h\infty}^{\mathrm{i}}$. $\mathbf{dev}(F^u)$ is a cone over $\mathbf{dev}(F_4^u)$, which must be a lune. $\mathbf{dev}(v_1^u)$, $\mathbf{dev}(v_2^u)$, and $\mathbf{dev}(v_3^u)$ are points in the boundary of $\mathbf{dev}(F_4^u)$. Since $\mathbf{dev}(F_4^i)$ converges to $\mathbf{dev}(F_4^u)$ geometrically, an elementary geometric consideration shows that two distinct points among $\mathbf{dev}(v_1^u)$, $\mathbf{dev}(v_2^u)$, and $\mathbf{dev}(v_3^u)$ form the vertices of the lune $\mathbf{dev}(F_4^u)$.

We claim that $v_1^u \neq v_2^u$. Otherwise, $\mathbf{dev}(v_1^i)$ and $\mathbf{dev}(v_2^i)$ converge to a common point respectively, and $\mathbf{dev}(G_1^i)$ converges to a segment since $\mathbf{dev}(G_1^i)$ is a cone over $\overline{\mathbf{dev}(v_1^i)\mathbf{dev}(v_5^i)}$ and $\overline{\mathbf{dev}(v_1^i)\mathbf{dev}(v_5^i)}$ is a subsegment of $\overline{\mathbf{dev}(v_1^i)\mathbf{dev}(v_2^i)}$. Since G_1^u is a nondegenerate 2-disk by Proposition 7.5, this is a contradiction. By Proposition 7.4, the sides F_1^u and F_2^u are not segments. Therefore, v_3^u is distinct from v_1^u and v_2^u. We obtain:

PROPOSITION 9.1. *The vertices v_1^u, v_2^u, and v_3^u are mutually distinct.*

So, it follows that v_1^u, v_2^u, v_3^u must be situated as in one of the following configurations (see Figure 9.1):

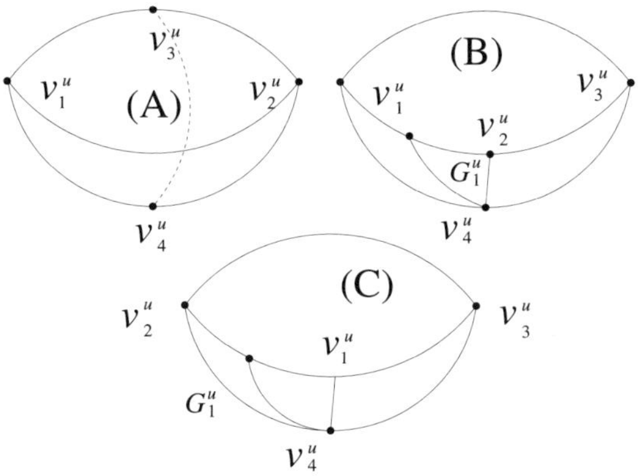

FIGURE 9.1. Cases A, B, C.

(A) v_1^u and v_2^u form antipodal vertices of the lune F_4^u with v_3^u in the interior of a segment connecting v_1^u and v_2^u.
(B) v_1^u and v_3^u form antipodal vertices of F_4^u with v_2^u in the interior of the segment connecting v_1^u and v_3^u.
(C) v_2^u and v_3^u form antipodal vertices of the lune F_4^u with v_1^u in the interior of the segment connecting v_2^u and v_3^u.

We will show that none of the above case is possible, starting with the case (A), using the strategy similar to what was in Chapter 8.

From our description (A), we see that v_5^u is a point of the segment I_{34}^u of **d**-length $= \pi$, v_3^u in the interior of the segment $I_{14}^u \cup I_{24}^u$ of **d**-length $= \pi$, and v_4^u at the origin O.

As in the above section, Proposition 7.4, Remark 6.2, Remark 7.1, and the endpoint matching condition show that we have only two possibilities:

(i) $F_1^u = U^u$, $F_2^u = D^u$, $L_1^u = \overline{v_3^u O}$, and $L_2^u = \overline{v_1^u v_3^u}$, or
(ii) $F_1^u = D^u$, $F_2^u = U^u$, $L_1^u = \overline{v_2^u v_3^u}$, and $L_2^u = \overline{v_3^u O}$.

(See Figures 8.2 and 8.3 also, which are still true schematically but not geometrically.)

From now on, $\overline{xyz}$ for three points x, y, z of $\mathbf{S}^3$ so that z is antipodal to x denotes the unique convex segment with endpoints x and z passing through y. Such a segment has **d**-length equal to π. Let T_s be the *standard trihedron* with vertices $[1,0,0,0]$ and $[-1,0,0,0]$ containing $[0,1,0,0]$ and $[0,0,1,0]$ and $[0,0,0,1]$ in its edges respectively. Now we choose projective automorphisms ψ and ψ^i so that

- $\psi(v_j)$ is in standard positions
$$[1,0,0,0], [0,1,0,0], [0,0,1,0], [0,0,0,1]$$
respectively for $j = 1, 2, 3, 4$,
- $\psi^i(\mathbf{dev}(v_j^i))$ are in standard positions for $j = 1, 3, 4$ respectively, and
- $\psi^i(\mathbf{dev}(I_{34}^i))$ is on the segment $\overline{[-1,0,0,0][0,1,0,0][1,0,0,0]}$ with endpoints $[1,0,0,0]$ and $\psi^i(\mathbf{dev}(v_2^i))$.

We choose ψ^i to form a bounded sequence in $\mathrm{Aut}(\mathbf{S}^3)$. (We assume that ψ^i converges to ψ^∞ in $\mathrm{Aut}(\mathbf{S}^3)$.) Since ψ^i is bounded, $\psi^i(\mathbf{dev}(v_2^i))$ converges to $[-1,0,0,0]$. Thus,

$$\psi_{\&}^i = \psi^i \circ h(\vartheta^i) \circ \psi^{-1} \tag{9.1}$$

acts on T_s and fixes the vertices of the standard trihedron T_s and $[0,0,1,0]$ and $[0,0,0,1]$; we let λ_k for $k = 1, 3, 4$ denote the eigenvalue of $\psi_{\&}^i$ associated with

$$[1,0,0,0], [0,0,1,0], \text{ and } [0,0,0,1]$$

respectively. $\phi_{\&}^i$ has the matrix form

$$\begin{bmatrix} \lambda_1^i & a^i & 0 & 0 \\ 0 & \lambda_2^i & 0 & 0 \\ 0 & 0 & \lambda_3^i & 0 \\ 0 & 0 & 0 & \lambda_4^i \end{bmatrix} \quad \text{for } a^i \in \mathbf{R}.$$

For convenience, we let $\psi(\mathbf{dev}(x_{12}))$, $\psi(\mathbf{dev}(x_{13}))$, and $\psi(\mathbf{dev}(x_{23}))$ equal $[0,0,1,1]$, $[0,1,0,1]$, and $[1,0,0,1]$ respectively. We also note that $\psi^\infty(\mathbf{dev}(F^u)) = \psi^\infty(F^\infty) = T_s$.

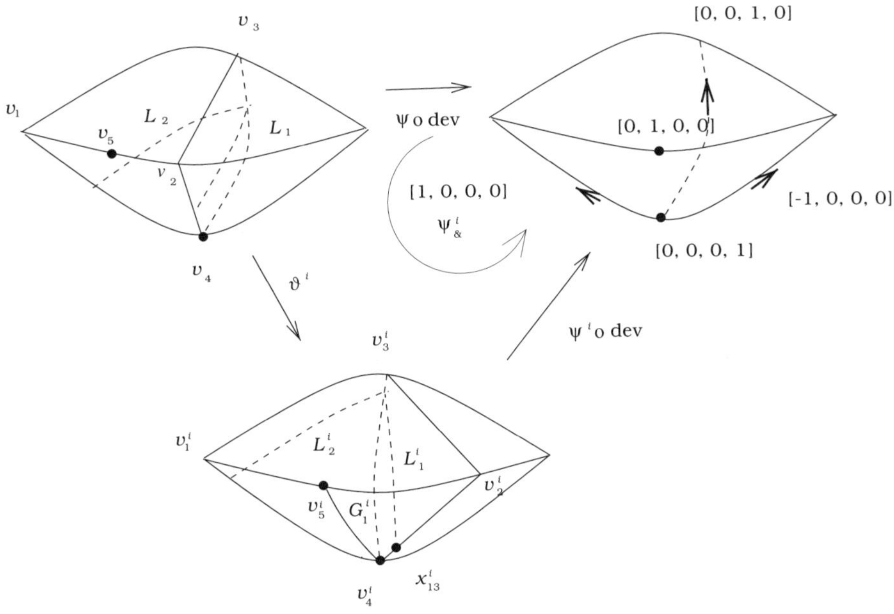

FIGURE 9.2. The case (A)(i).

(i) Notice that $\psi^i_\&$ is acting on the lune B with vertices $[1,0,0,0]$ and $[-1,0,0,0]$ and containing $[0,1,0,0]$ and $[0,0,0,1]$ in its edges where $\psi^i_\&([0,1,0,0])$ converges to $[-1,0,0,0]$ and $\psi^i(\mathbf{dev}(x^i_{13}))$ converges to $[0,0,0,1]$.

From the description of the endpoints of L_1 and L_2, Table 1 on x_{23} and x_{12} and Lemma 8.1, we obtain that $\lambda^i_1, \lambda^i_3 \gg \lambda^i_4$.

In the affine 3-space $\mathcal{H}^o$ with origin $[0,0,0,1]$, $\psi^i_\&$ is a linear map represented by a 3×3-matrix.

We introduce the coordinates on the affine plane containing B^o so that $[0,0,0,1]$ has coordinates $(0,0)$ and $\overline{[1,0,0,0][0,0,0,1]} = \psi(I_{23})$ and $\overline{[0,1,0,0][0,0,0,1]} = \psi(I_{13})$ correspond to the x-axis and the y-axis respectively. $\psi(\mathbf{dev}(x_{13}))$ and $\psi(\mathbf{dev}(x_{23}))$ have coordinates $(1,0)$ and $(0,1)$ respectively.

If we restrict $\psi^i_\&$ to the vector subspace corresponding to the lune B, then $\psi^i_\&$ has a matrix expression

$$M^i_1 = \begin{bmatrix} a_i & b_i \\ 0 & d_i \end{bmatrix}, a_i, d_i > 0.$$

Since $\psi^i(\mathbf{dev}(x^i_{23})) \to [1,0,0,0]$, by equation 9.1 we have $a_i \to \infty$, and since $\psi^i(\mathbf{dev}(x^i_{13})) \to [0,0,0,1]$, we obtain $b_i, d_i \to 0$. Since $[0,0,1,0]$ is a fixed point of $\psi^i_\&$, it follows that $\psi^i_\&$ has a 3×3-matrix expression:

$$M^i_2 = \begin{bmatrix} a_i & b_i & 0 \\ 0 & d_i & 0 \\ 0 & 0 & e_i \end{bmatrix} \quad a_i, b_i, e_i > 0.$$

The eigenvalues of M^i_1 are a_i and d_i and the corresponding eigenvectors are $(1,0)$ and $(-b_i/(a_i-d_i), 1)$. Since $b_i, d_i \to 0$ and $a_i \to \infty$, it follows that the sequence of the second eigenvectors converges to $(0,1)$. Thus, for sufficiently large i, the matrix M^i_2 has a fixed point s_i near $[0,1,0,0]$ on the segment $\overline{[1,0,0,0][0,1,0,0][-1,0,0,0]}$, with the associated eigenvalue d_i. Since $a_i, e_i \to \infty$ as $\lambda_1^i, \lambda_3^i \gg \lambda_4^i$, and $d_i \to 0$, it follows that d_i is the least eigenvalue sequence of M_2^i.

Choose a point x on the edge I_{13} of F and a tiny-ball neighborhood $B(x)$ of x and form a radiant cone C_x containing $B(x)$. By the condition on the eigenvalues and eigenvectors of M_2^i, $\psi_{\&}^i(\psi(\mathbf{dev}(\mathrm{Cl}(C_x))))$ converges to a radiant bihedron including T_s. The trihedron T_s equals $\psi^\infty(\mathbf{dev}(F^u))$. Since $\psi^i \circ \mathbf{dev}(\vartheta^i(\mathrm{Cl}(C_x)))$ converges to a radiant bihedron including T_s, $\mathbf{dev}(\vartheta^i(\mathrm{Cl}(C_x)))$ converges to a radiant bihedron including F^∞. As in the case (i) of Chapter 8, since $\vartheta^i(\mathrm{Cl}(C_x)))$ always overlaps $\vartheta^i(F)$, the dominating part of Theorem B.1 shows that $\check{M}_h$ includes a radiant bihedron, contradicting Hypothesis 1.

(ii) Here $L_1^u = \overline{v_2^u v_3^u}$ and $L_2^u = \overline{v_3^u v_4^u}$. Since $x_{23}^u = O$ and $x_{12}^u = v_3^u$, we have that $\lambda_3 \gg \lambda_4 \gg \lambda_1$ by Table 1 and Lemma 8.1.

Our transformation $\psi_{\&}^i$ acts on the lune B with vertices $[1,0,0,0]$ and $[-1,0,0,0]$ and containing $[0,1,0,0]$ and $[0,0,0,1]$ in its edges. We introduce coordinates as above and let M_1^i denote the 2×2-matrix corresponding to $\psi_{\&}^i$ acting on the vector subspace corresponding to B:

$$M_1^i = \begin{bmatrix} a_i & b_i \\ 0 & d_i \end{bmatrix} \quad a_i, d_i > 0.$$

Since $\psi^i(\mathbf{dev}(x_{23}^i))$ converges to $[0,0,0,1]$, we obtain $a_i \to 0$. Since the direction of $\overline{O\psi^i(\mathbf{dev}(x_{13}))}$ converges to that of $\overline{O[-1,0,0,0]}$, it follows that $b_i/d_i \to -\infty$. Here $a_i, d_i > 0$ and $b_i < 0$ for i sufficiently large. Moreover, $b_i \to -\infty$ by the condition that $\psi^i(\mathbf{dev}(x_{13}^i))$ converges to $[-1,0,0,0]$.

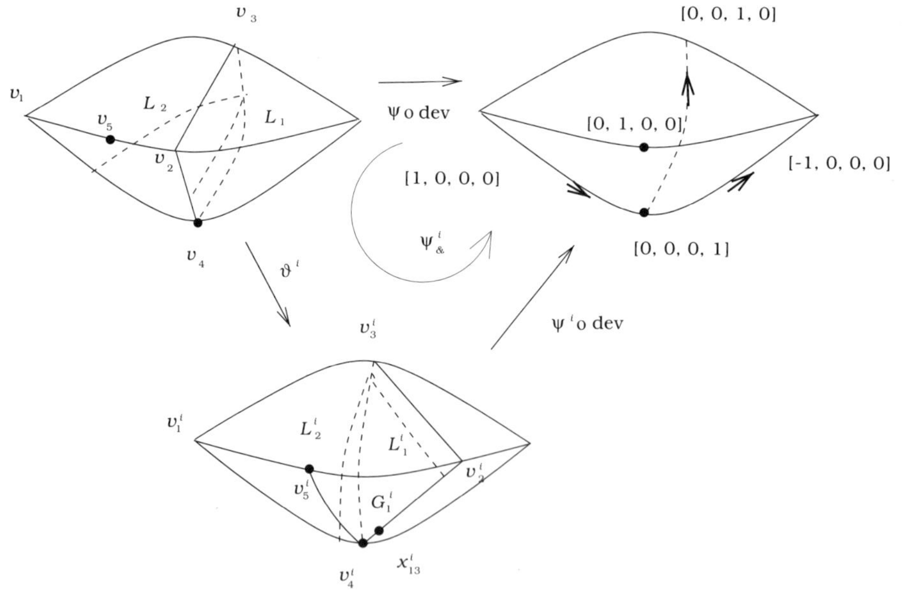

FIGURE 9.3. The case (A)(ii).

For any pair (c, d) of numbers, $d > 0$, M_1^i maps (c, d) to $(ca_i + db_i, dd_i)$. Since we have

$$\left|\frac{c}{d}\frac{a_i}{d_i}\right| / \left|\frac{b_i}{d_i}\right| = \left|\frac{c}{d}\right| \left|\frac{a_i}{b_i}\right| \to 0,$$

the ratio of the coordinates $(ca_i + db_i)/(dd_i) = (c/d)a_i/d_i + b_i/d_i$ converges to $-\infty$. Thus, for every point x of B^o, the sequence of rays $\psi_{\&}^i(\overline{Ox})$ converges to $\overline{O[-1,0,0,0]}$, meaning that $\psi_{\&}^i \circ \psi(\mathbf{dev}(\mathrm{Cl}(G_1)))$ converges to $\overline{O[-1,0,0,0]}$. This shows that $h(\vartheta^i)(\mathbf{dev}(\mathrm{Cl}(G_1)))$ converges to a segment, contradicting Proposition 7.5.

We now rule out the case (B). In this case, F_2^u is a lune opposite to v_2^u and F_1^u is a triangle, and so is F_3^u. As in case (A), Proposition 7.4, Remarks 7.1 and 6.2, and the endpoint matching condition show that

(i) $F_1^u = D^u$, $F_2^u = U^u$, L_1^u is the segment I_{14}^u and L_2^u is a subsegment of $I_{12}^u \cup I_{23}^u$ with endpoints v_3^u and $x_{23}^u \in I_{23}^u$.

(ii) $F_1^u = U^u$, $F_2^u = D^u$, L_1^u equals I_{12}^u, and L_2^u equals I_{24}^u.

By the convergence condition, i.e., equation (6.2), we may find a unique element ψ of $\mathrm{Aut}(\mathbf{S}^n)$ and a sequence of uniformly bounded transformations ψ^i of $\mathrm{Aut}(\mathbf{S}^n)$ such that

- $\psi(\mathbf{dev}(v_1)), \psi(\mathbf{dev}(v_2)), \psi(\mathbf{dev}(v_3))$, and $\psi(\mathbf{dev}(v_4))$ are at

$$[1,0,0,0], [0,1,0,0], [0,0,1,0], \text{ and } [0,0,0,1]$$

  respectively,
- $\psi^i(\mathbf{dev}(v_1^i)) = [1,0,0,0]$, $\psi^i(\mathbf{dev}(v_2^i)) = [0,1,0,0]$, and $\psi^i(\mathbf{dev}(v_4^i)) = [0,0,0,1]$.
- $\psi^i(\mathbf{dev}(I_{24}^i))$ is a segment in $\overline{[1,0,0,0][0,0,1,0][-1,0,0,0]}$ with endpoints $[1,0,0,0]$ and $\psi^i \circ \mathbf{dev}(v_3^i)$, the latter of which forms a sequence converging to $[-1,0,0,0]$.

We assume as before that ψ^i converges to ψ^∞ in $\mathrm{Aut}(\mathbf{S}^3)$.

As in the previous chapters, we define $\psi_{\&}^i$ to be $\psi^i \circ h(\vartheta^i) \circ \psi^{-1}$. Then $\psi_{\&}^i$ acts on T_s and fixes each of $[1,0,0,0], [0,1,0,0], [-1,0,0,0], [0,0,0,1]$ but not $[0,0,1,0]$. Thus, $\psi_{\&}^i$ has a matrix form:

$$\begin{bmatrix} \lambda_1^i & 0 & a^i & 0 \\ 0 & \lambda_2^i & 0 & 0 \\ 0 & 0 & \lambda_3^i & 0 \\ 0 & 0 & 0 & \lambda_4^i \end{bmatrix} \tag{9.2}$$

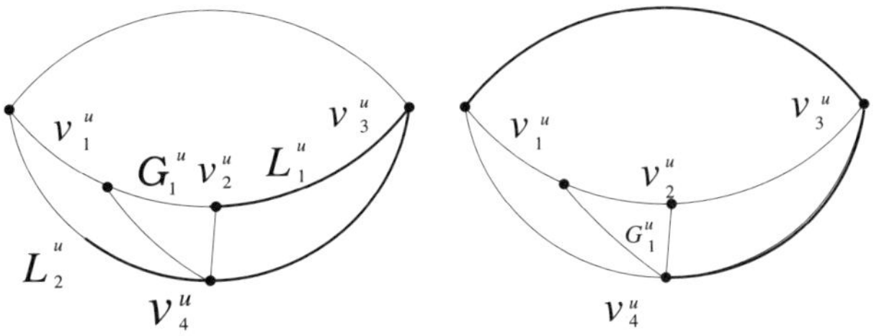

FIGURE 9.4. Cases (B)(i) and (B)(ii).

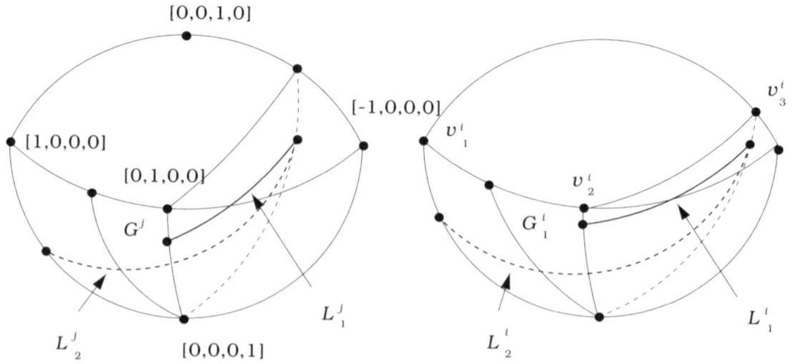

FIGURE 9.5. The case (B)(i).

Obviously, we may assume $\lambda_1^i, \lambda_2^i, \lambda_4^i > 0$. As we require $\lambda_1^i\lambda_2^i\lambda_3^i\lambda_4^i = 1$, it follows that $\lambda_3^i > 0$.

For convenience, we assume that $\psi(\mathbf{dev}(x_{12}))$, $\psi(\mathbf{dev}(x_{23}))$, and $\psi(\mathbf{dev}(x_{13}))$ equal $[0,0,1,1]$, $[1,0,0,1]$, and $[0,1,0,1]$ respectively. Again, $\psi^\infty(\mathbf{dev}(F^u)) = T_s$.

We begin with (B)(i). By condition (i), we see that $\psi^i \circ \mathbf{dev}(x_{13}^i)$ converges to $[0,1,0,0]$, $\psi^i \circ \mathbf{dev}(x_{12}^i)$ to $[-1,0,0,0]$, and $\psi^i \circ \mathbf{dev}(L_1^i)$ converges to the segment $\overline{[0,1,0,0][-1,0,0,0]}$. $\psi^i \circ \mathbf{dev}(x_{23}^i)$ converges to a point of $\overline{[1,0,0,0][-1,0,0,0]}$, and $\psi^i \circ \mathbf{dev}(L_2^i)$ converges to a subsegment of $\overline{[1,0,0,0][0,0,0,1][-1,0,0,0]}$ with an endpoint $[-1,0,0,0]$ and containing $[0,0,0,1]$. (See Figure 9.5.)

From Table 1, we will get estimations of eigenvalues of $\psi_\&^i$ using Lemma 8.1. By considering x_{13}^u, we obtain $\lambda_2^i \gg \lambda_4^i$, and from the following lemma we get $\lambda_1^i \gg \lambda_4^i$: Hence, $\psi_\&^i \circ \psi \circ \mathbf{dev}(x_{23})$ converges to $[1,0,0,0]$ and so does $\psi^i \circ \mathbf{dev}(x_{23}^i)$; $\psi^i \circ \mathbf{dev}(L_2^i)$ converges to $\overline{[1,0,0,0][0,0,0,1][-1,0,0,0]}$.

LEMMA 9.1. *$\lambda_1^i \gg \lambda_2^i$ or $\lambda_1^i \sim \lambda_2^i$.*

PROOF. Since $\psi^i \circ \mathbf{dev}(F_3^i)$ equals the triangle

$$T = \triangle([1,0,0,0][0,1,0,0][0,0,0,1]),$$

$\psi_\&^i$ acts on T. If $\lambda_2^i \gg \lambda_1^i$, then $\psi_\&^i \circ \psi(\mathbf{dev}(\mathrm{Cl}(G_1))) = \psi^i(\mathbf{dev}(G_1^i))$ converges to the segment $\overline{[0,1,0,0][0,0,0,1]}$. This contradicts Proposition 7.5. □

Let us denote by B the lune with vertices $[1,0,0,0]$ and $[-1,0,0,0]$ and two sides passing through $[0,0,0,1]$ and $[0,0,1,0]$. Let $\mathbf{S}_B^2$ denote the great sphere including B. Then $\psi_\&^i$ restricts to a projective automorphism on $\mathbf{S}_B^2$ with the matrix form

$$A_i = \begin{bmatrix} \lambda_1^i & 0 & a^i \\ 0 & \lambda_4^i & 0 \\ 0 & 0 & \lambda_3^i \end{bmatrix} \tag{9.3}$$

where $[1,0,0,0]$, $[0,0,0,1]$, and $[0,0,1,0]$ form the order of the coordinate system.

A *dilatation* is defined to be a projective automorphism of $\mathbf{S}^3$ with a matrix of form

$$A_\lambda = \begin{bmatrix} 1 & 0 & 0 & 0 \\ 0 & 1 & 0 & 0 \\ 0 & 0 & 1 & 0 \\ 0 & 0 & 0 & \lambda \end{bmatrix}, \lambda > 0 \tag{9.4}$$

with respect to the standard basis $[1,0,0,0], [0,1,0,0], [0,0,1,0]$, and $[0,0,0,1]$

LEMMA 9.2. *Let ϕ^i be a sequence of projective maps acting on the standard bihedron T_s. Let S be a segment with vertices respectively at the middle of two radial edges e and f of a standard tetrahedron. Suppose that $\phi^i(S)$ converges to a compact set S^*. Assume that S^* is dilatation-invariant. If S' is another segment with vertices in e^o and f^o. Then $\phi^i(S')$ also converges to S^*.*

PROOF. Suppose that S' equals $A(S)$ for a dilatation A. Then as A commutes with ϕ^i, $\phi^i(A(S)) = A(\phi^i(S))$ converges to S^* for each λ.

Suppose now that S' is arbitrary. By choosing $\lambda \gg 1$ and $\delta \ll 1$, we can show that our S' lies in the quadrilateral in a side of a standard tetrahedron bounded by $A_\lambda(S)$ and $A_\delta(S)$ and segments in e and f respectively. Since both sequences of images of $A_\lambda(S)$ and $A_\delta(S)$ under ϕ^i converges to S^*, it follows that $\phi^i(S')$ converges to S^*. □

Recall that $\psi^i_\&(\psi(\mathbf{dev}(L_2)))$ converges to

$$\overline{[1,0,0,0][0,0,0,1][-1,0,0,0]},$$

a dilatation-invariant set. This is true no matter how we chose x_{ij} in I_{ij}: Let x'_{ij} denote another choice in I_{ij}. Let L'_1 and L'_2 be the resulting segments for our new choice. By above lemma 9.2, the limit of the sequence of images $\psi(\mathbf{dev}(L'_2))$ under $\psi^i_\&$ is same as that of $\psi(\mathbf{dev}(L_2))$. (For L_1 and L'_1, we can say similar things.)

On $\mathbf{S}^2_\infty$, $\psi^i_\&$ restricts to a projective map with the matrix form

$$B_i = \begin{bmatrix} \lambda^i_1 & 0 & a^i \\ 0 & \lambda^i_2 & 0 \\ 0 & 0 & \lambda^i_3 \end{bmatrix} \tag{9.5}$$

where $[1,0,0,0]$, $[0,1,0,0]$, and $[0,0,1,0]$ form the order of the basis.

Since $\lambda^i_2 \gg \lambda^i_4$, we see that $B_i = K_i A_i$ (strictly as 3×3-matrices) for K_i equal to

$$\begin{bmatrix} 1 & 0 & 0 \\ 0 & k_i & 0 \\ 0 & 0 & 1 \end{bmatrix} \tag{9.6}$$

where k_i converges to ∞. Let L''_2 be the segment in F_4 with vertices in the interiors of I_{14} and I_{34} respectively. Then by above equation, we easily see that the limit of the sequence $\psi^i_\&(L''_2)$ is a subsegment of

$$\overline{[1,0,0,0][0,1,0,0][-1,0,0,0]}$$

or the singleton $\{[0,1,0,0]\}$ up to choosing a subsequence. This follows since under the sequence of projective automorphisms of $\mathbf{S}^2_\infty$ with *matrices* A_i, this happens already for L''_2 by the conclusion of the above paragraphs and K_i only enforces this for L''_2 to make the limit to be as stated since K_i pushes everything near $[0,1,0,0]$. (We are applying facts obtained for $\mathbf{S}^B$ to $\mathbf{S}^2_\infty$ in a somewhat convoluted way.)

A *cone-neighborhood* of a subset A of $\mathcal{H}$ is a radial-flow invariant neighborhood of A in $\mathcal{H}$.

The triangle $\psi \circ \mathbf{dev}(F_2)$ has vertices $[1,0,0,0], [0,0,1,0]$, and $[0,0,0,1]$. Consider a cone-neighborhood of $\psi \circ \mathbf{dev}(F_2 \cap M_h) - \{O\}$ and a quadrilateral disk J in $\mathbf{S}^2_\infty$ with vertices the vertices of L''_2 and $[1,0,0,0]$ and $[0,0,1,0]$. Reflect this quadrilateral disk by an order two **d**-isometric reflection R_2 fixing points of the triangle $\psi \circ \mathbf{dev}(F_2)$. Then the union of the quadrilateral disk and its reflected image is another quadrilateral disk, say J'. If we choose the vertices of L''_2 sufficiently close to the edge $\overline{[1,0,0,0][0,0,1,0]}$, the interior of the cone over J' is a subset of the cone-neighborhood of $\psi \circ \mathbf{dev}(F_2 \cap M_h) - \{O\}$.

Since $F_2 \cap M_h$ is a subset of M_h, there exists a cone-neighborhood C of $F_2 \cap M_h$ mapping into the above cone-neighborhood of $\phi \circ \mathbf{dev}(F_2) - \{O\}$ under $\phi \circ \mathbf{dev}$ if we choose the cone-neighborhood sufficiently close to $\psi \circ \mathbf{dev}(F) - \{O\}$. Therefore, $\mathrm{Cl}(C)$ includes a compact 3-ball J'' mapping to the cone over J' under $\phi \circ \mathbf{dev}$.

By the condition on the limit of the sequence of images of L''_2 under $\psi^i_\&$, we can show that $\psi^i_\&(J)$ converges to the lune in $\mathbf{S}^2_\infty$ with vertices $[1,0,0,0]$ and $[-1,0,0,0]$ with edges passing through $[0,1,0,0]$ and $[0,0,1,0]$ respectively. As R_2 commutes with $\psi^i_\&$, we see that the limit of $\psi^i_\&(J')$ is a 2-hemisphere with center $[0,0,1,0]$. Therefore, $\psi^i_\&(\psi \circ \mathbf{dev}(J''))$ converges to a radiant bihedron including T_s. The trihedron T_s equals $\psi^\infty(\mathbf{dev}(F^u)) = \psi^\infty(F^\infty)$ the limit of $\psi^i_\&(\psi \circ \mathbf{dev}(F))$. Thus, $\mathbf{dev}(\vartheta^i(J''))$ converges to a radiant bihedron including F^∞ up to a choice of subsequences. As $\vartheta^i(J'')$ always overlaps with $\vartheta^i(F)$, similarly to case (i) in Chapter 8 the dominating part of Theorem B.1 shows that there exists a radiant bihedron in $\check{M}_h$, a contradiction.

We will now study (B)(ii). By condition (B)(ii), we see that $\psi^i \circ \mathbf{dev}(x^i_{23})$ converges to $[1,0,0,0]$, $\psi^i \circ \mathbf{dev}(x^i_{13})$ to $[0,0,0,1]$, $\psi^i \circ \mathbf{dev}(x^i_{12})$ to $[-1,0,0,0]$, and $\psi^i \circ \mathbf{dev}(L^i_1)$ to the segment $\overline{[0,0,0,1][-1,0,0,0]}$. $\psi^i \circ \mathbf{dev}(L^i_2)$ converges to the segment $\overline{[1,0,0,0][0,0,1,0][-1,0,0,0]}$.

From the condition on x^i_{13}, we have that $\lambda^i_4 \gg \lambda^i_2$. By the condition on x^i_{23}, we get $\lambda^i_1 \gg \lambda^i_4$, and obtain

$$\lambda^i_1 \gg \lambda^i_4 \gg \lambda^i_2. \tag{9.7}$$

As above, we denote by B the lune with vertices $[1,0,0,0]$ and $[-1,0,0,0]$ and two sides passing through $[0,0,0,1]$ and $[0,0,1,0]$.

Let $\mathbf{S}^2_B$ denote the great sphere including B. Then $\psi^i_\&$ restricts to a projective automorphism on $\mathbf{S}^2_B$ with the matrix form

$$A_i = \begin{bmatrix} \lambda^i_1 & 0 & a^i \\ 0 & \lambda^i_4 & 0 \\ 0 & 0 & \lambda^i_3 \end{bmatrix} \tag{9.8}$$

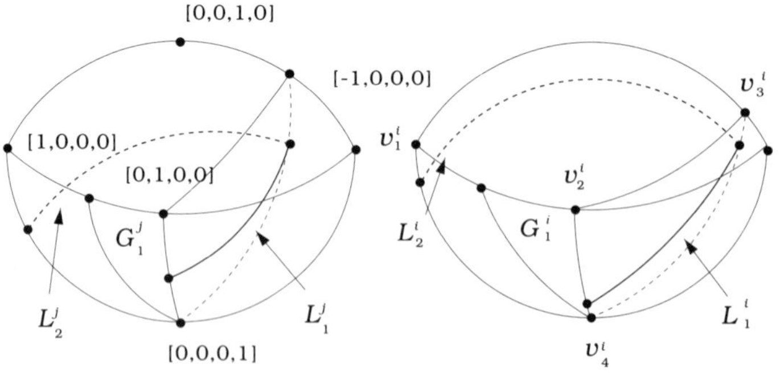

FIGURE 9.6. The case (B)(ii).

where $[1,0,0,0]$, $[0,0,0,1]$, and $[0,0,1,0]$ form the order of the coordinate system.

By assumption, $\psi^i_\&(\psi(\mathbf{dev}(L_2)))$ converges to

$$\overline{[1,0,0,0][0,0,1,0][-1,0,0,0]}.$$

This is true no matter how we chose x_{ij} in I_{ij} as above by Lemma 9.2.

On $\mathbf{S}^2_\infty$, $\psi^i_\&$ restricts to a projective map with the matrix form

$$B_i = \begin{bmatrix} \lambda^i_1 & 0 & a^i \\ 0 & \lambda^i_2 & 0 \\ 0 & 0 & \lambda^i_3 \end{bmatrix}. \tag{9.9}$$

Since $\lambda^i_4 \gg \lambda^i_2$, we see that $B_i = K_i A_i$ for K_i equal to

$$\begin{bmatrix} 1 & 0 & 0 \\ 0 & k_i & 0 \\ 0 & 0 & 1 \end{bmatrix} \tag{9.10}$$

where k_i converges to 0. Let L''_2 be the segment in F_4 with vertices in the interiors of I_{14} and I_{34} respectively. A sequence of segments obtained by projective automorphisms of $\mathbf{S}^2_\infty$ with *matrices* A_i applied to (L''_2) converges to

$$\overline{[1,0,0,0][0,0,1,0][-1,0,0,0]},$$

a fact following from seeing what happens in $\mathbf{S}^2_B$ under $\psi^i_\&$. Then by the above equation, we easily see that the limit of $\psi^i_\&(L''_2)$ is the segment

$$\overline{[1,0,0,0][0,0,1,0][-1,0,0,0]}$$

as K_i can only help the segments to do so.

Let L''_1 be the segment in F_4 with endpoints in endpoints in the interior of I_{14} and $\overline{[0,1,0,0][-1,0,0,0]}$ and the endpoint in I_{14} agrees with that of L''_2, which we denote by x''_{14}. Then the limit of the sequence of images of x''_{14} under $\psi^i_\&$ is clearly $[-1,0,0,0]$. Let y be the other endpoint of L''_1. Since $\lambda^i_1 \gg \lambda^i_2$, we see that the limit of the sequence of images of y under $\psi^i_\&$ equals $[-1,0,0,0]$. Thus, that of L''_1 under $\psi^i_\&$ converges to $\{[-1,0,0,0]\}$.

Choose a cone-neighborhood C of I^o_{13} in M_h. We choose L''_1 and L''_2 sufficiently close to $[0,1,0,0]$ so that they lie in $\phi\circ\mathbf{dev}(\mathrm{Cl}(C))$. Let R be the $\mathbf{d}$-reflection so that the set of fixed points is precisely the great circle containing $[0,1,0,0]$ and $[0,0,0,1]$. Then R commutes with $\psi^i_\&$. L''_1, L''_2, $R(L''_1)$, and $R(L''_2)$ bound a quadrilateral disk J' in $\mathbf{S}^2_\infty$. We of course assume that $R(L''_1)$ and $R(L''_2)$ also lies in $\phi\circ\mathbf{dev}(\mathrm{Cl}(C))$.

We see easily that the limit of the sequence of the image of J' under $\psi^i_\&$ converges to the hemisphere with center $[0,1,0,0]$. There exists a cone in J'' in $\mathrm{Cl}(C)$ which maps to the cone over J' under $\psi\circ\mathbf{dev}$. The sequence of images of J'' under $\psi^i_\&\circ\psi\circ\mathbf{dev} = \psi^i\circ\mathbf{dev}\circ\vartheta^i$ converges to a radiant bihedron including $\psi^\infty(\mathbf{dev}(F^u)) = T_s$ by the above discussion. That of images of J'' under $\mathbf{dev}\circ\vartheta^i$ converges to a radiant bihedron including F^∞. As above by Theorem B.1(4), there exists a radiant bihedron in $\check{M}_h$, a contradiction to Hypothesis 1. We showed (B)(ii) does not happen.

Now we will show that (C) does not happen. Since the arguments are similar to the case (B), we will only sketch the argument:

In this case, F^u_1 is a lune opposite v^u_1 and F^u_2 is a triangle, and so is F^u_3. As in case (A) or (B), one of the following is true:

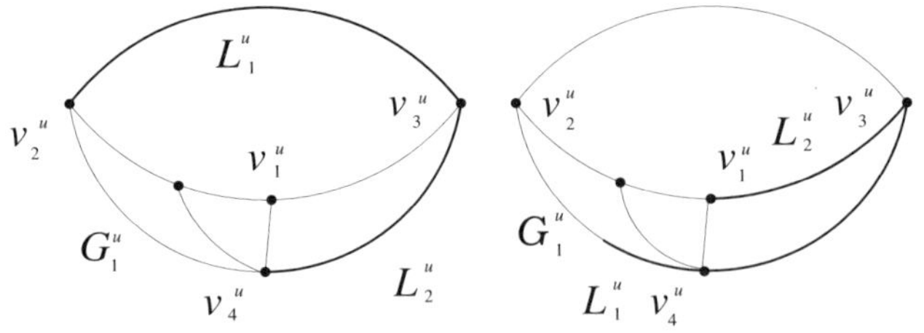

FIGURE 9.7. Cases (C)(i) and (C)(ii).

(i) We have $F_1^u = D^u$, $F_2^u = U^u$, and L_1^u is the segment I_{14}^u and L_2^u is a segment I_{12}^u.
(ii) We have $F_1^u = U^u$, $F_2^u = D^u$, and L_1^u is a subsegment of $I_{12}^u \cup I_{13}^u$ with an endpoint v_3^u and x_{13}^u and L_2^u equals I_{24}^u.

By the convergence condition equation 6.2, we may find a unique element ψ of $\mathrm{Aut}(\mathbf{S}^3)$ and a sequence of uniformly bounded transformations ψ^i of $\mathrm{Aut}(\mathbf{S}^3)$ such that

- $\psi(\mathbf{dev}(v_1)), \psi(\mathbf{dev}(v_2)), \psi(\mathbf{dev}(v_3))$, and $\psi(\mathbf{dev}(v_4))$ are at
$$[0,1,0,0], [1,0,0,0], [0,0,1,0], \text{ and } [0,0,0,1]$$
respectively,
- $\psi^i(\mathbf{dev}(v_1^i)) = [0,1,0,0]$, $\psi^i(\mathbf{dev}(v_2^i)) = [1,0,0,0]$, $\psi^i(\mathbf{dev}(v_4^i)) = [0,0,0,1]$, and
- $\psi^i(\mathbf{dev}(I_{14}^i))$ is a segment in $\overline{[1,0,0,0][0,0,1,0][-1,0,0,0]}$, with endpoints $[1,0,0,0]$ and $\psi^i \circ \mathbf{dev}(v_3^i)$, the latter of which forms a sequence converging to $[-1,0,0,0]$.

As before, we define $\psi_{\&}^i$ to be $\psi^i \circ h(\vartheta^i) \circ \psi^{-1}$. Then $\psi_{\&}^i$ fixes each of $[1,0,0,0]$, $[0,1,0,0]$, $[-1,0,0,0]$, and $[0,0,0,1]$, and has a matrix form:

$$\begin{bmatrix} \lambda_1^i & 0 & a^i & 0 \\ 0 & \lambda_2^i & 0 & 0 \\ 0 & 0 & \lambda_3^i & 0 \\ 0 & 0 & 0 & \lambda_4^i \end{bmatrix} \tag{9.11}$$

where λ_j^i are positive as before.

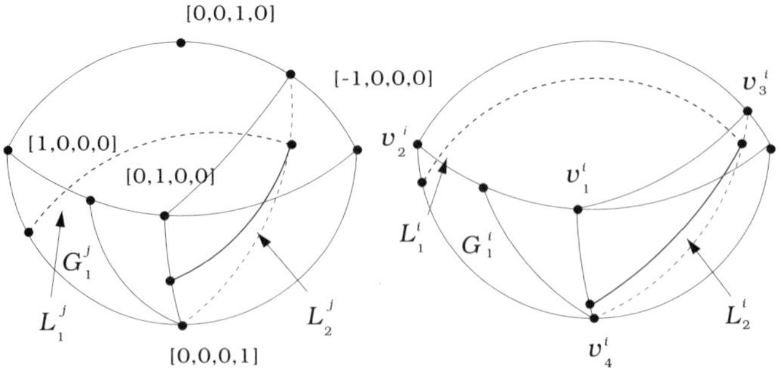

FIGURE 9.8. The case (C)(i).

We begin with (C)(i). By condition (i), we see that $\psi^i \circ \mathbf{dev}(x^i_{23})$ converges to $[0,0,0,1]$, $\psi^i \circ \mathbf{dev}(x^i_{13})$ to $[1,0,0,0]$, and hence $\psi^i \circ \mathbf{dev}(L^i_1)$ to the segment

$$\overline{[1,0,0,0][0,0,0,1][-1,0,0,0]}.$$

From this, we obtain the eigenvalue estimation $\lambda^i_4 \gg \lambda^i_2$ and $\lambda^i_1 \gg \lambda^i_4$; which implies $\lambda^i_1 \gg \lambda^i_2$. Then $\psi^i_{\&}(\psi \circ \mathbf{dev}(G^u_1)) = \psi^i \circ \mathbf{dev}(\vartheta^i(G^u_1))$ converges to the segment $\overline{[0,0,0,1][0,1,0,0]}$. As ψ^i is bounded, this is a contradiction to Proposition 7.5.

We finish the series of the arguments with (C)(ii). By condition (ii), $\psi^i \circ \mathbf{dev}(x^i_{23})$ converges to $[0,1,0,0]$, implying $\lambda^i_2 \gg \lambda^i_4$.

As above, we denote by B the lune with vertices $[1,0,0,0]$ and $[-1,0,0,0]$ and two sides passing through $[0,0,0,1]$ and $[0,0,1,0]$.

Let $\mathbf{S}^2_B$ denote the great sphere including B. Then $\psi^i_{\&}$ restricts to a projective automorphism on $\mathbf{S}^2_B$ with the matrix form

$$A_i = \begin{bmatrix} \lambda^i_1 & 0 & a^i \\ 0 & \lambda^i_4 & 0 \\ 0 & 0 & \lambda^i_3 \end{bmatrix} \tag{9.12}$$

where $[1,0,0,0]$, $[0,0,0,1]$, and $[0,0,1,0]$ form the order of the coordinate system.

By assumption, $\psi^i_{\&}(\psi(\mathbf{dev}(L_1)))$ converges to a subsegment of

$$\overline{[1,0,0,0][0,0,0,1][-1,0,0,0]}.$$

This is true no matter how we chose x_{ij} in I_{ij} as before.

On $\mathbf{S}^2_\infty$, $\psi^i_{\&}$ restricts to a projective map with the matrix form

$$B_i = \begin{bmatrix} \lambda^i_1 & 0 & a^i \\ 0 & \lambda^i_2 & 0 \\ 0 & 0 & \lambda^i_3 \end{bmatrix}. \tag{9.13}$$

Since $\lambda^i_2 \gg \lambda^i_4$, we see that $B_i = K_i A_i$ for K_i equal to

$$\begin{bmatrix} 1 & 0 & 0 \\ 0 & k_i & 0 \\ 0 & 0 & 1 \end{bmatrix} \tag{9.14}$$

where k_i converges to ∞. Let L''_1 be the segment in F_4 with vertices in the interiors of I_{14} and I_{34} respectively. Then as in case (B)(i), we easily see that the limit of the sequence $\psi^i_{\&}(L''_1)$ is a subsegment of $\overline{[1,0,0,0][0,1,0,0][-1,0,0,0]}$ or the set of the point $[0,1,0,0]$.

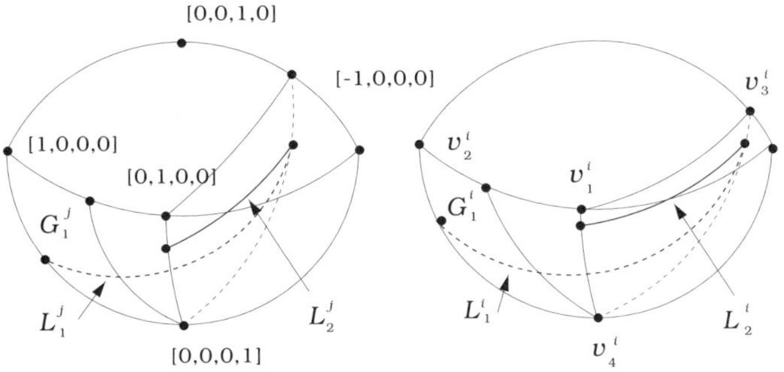

FIGURE 9.9. The case (C)(ii).

Exactly the same steps in (B)(i) will apply from here on and show that there exists a radiant bihedron in $\check{M}_h$, which is a contradiction. Hence, the proof of Proposition 7.1 is completed.

CHAPTER 10

Obtaining concave-cone affine manifolds

We will discuss the transversal intersection of two crescent-cones, and define the union $\Lambda^c(R)$ of a collection of overlapping crescent-cones for a crescent-cone R in $\check{M}_h$. The main part of the chapter is to derive the properties of $\Lambda^c(R)$, particularly, the equivariance and local finiteness. We may find "two-faced manifolds" as in n-crescent cases in [**15**]. After splitting off along these totally geodesic surfaces, we will be able to show that the union of all subsets of form $\Lambda^c(R)$ for a crescent-cone R covers a compact submanifold in of codimension 0, i.e., the union of concave-cone affine submanifolds, using Proposition 2.1. The complement of this compact manifold is convex, and hence, we will achieve the decomposition. This will in turn prove Theorem A, and Corollaries A and B.

The main result of this chapter, implying Theorem A, is as follows:

THEOREM 10.1. *Assume that M is a 2-convex but nonconvex radiant affine compact 3-manifold with totally geodesic or empty boundary. The Kuiper completion of the holonomy cover of M does not include a radiant bihedron. Then M decomposes along disjoint totally geodesic tori or Klein bottles into convex radiant affine 3-manifolds and concave-cone affine manifolds.*

We will be splitting and decomposing M in the following steps. The completions of the holonomy cover of the resulting manifolds also do not include radiant bihedra and pseudo-crescent-cones. We need this to insure that the constructed manifold in each step satisfies the required assumptions for the next step.

PROPOSITION 10.1. *Let N be a radiant affine 3-submanifold of M, closed in M or a radiant affine 3-manifold obtained from M by splitting along a properly imbedded totally geodesic surfaces. Then if the completion $\check{N}_h$ of the holonomy cover N_h of N includes a radiant bihedron, a crescent-cone, or a pseudo-crescent cone, then so does $\check{M}_h$.*

PROOF. This is similar to the proof of Proposition 8.9 of [**15**]. Basically, we show that N_h is a subset of M_h when N is a submanifold of M, and if R is a radiant bihedron in $\check{N}_h$, then the closure of R^o in M_h is a radiant bihedron in $\check{M}_h$. The same argument works if R is a crescent-cone or a pseudo-crescent-cone. The details are omitted since they are more completely explained in [**15**]. When N is obtained from M by splitting, a similar argument also applies. □

From now on, we assume that $\check{M}_h$ includes crescent-cones but no pseudo-crescent-cones, which will be proved in Chapter 12. We also assume that $\check{M}_h$ does not include any radiant bihedron because we decomposed using results of [**15**] (see Theorem 4.1).

We will now construct an object in $\check{M}_h$ playing an analogous role to concave set in [**11**]; that is, we will find an equivariant and locally finite set from the suitable collections of crescent-cones.

For a convex polyhedron or polygon G in $\check{M}_h$, a side meeting M_h is said to be a *finite side* of F. A side of G in $M_{h\infty}$ is said to be an *ideal side*. If it is in $M^{\mathsf{i}}_{h\infty}$, we call it an *infinitely ideal side*, and if not, a *finitely ideal side*.

DEFINITION 10.1. Let T be a crescent-cone in $\check{M}_h$, which is a trihedron with three lune sides. We denote by ν_T the side of T meeting M_h, by α^{f}_T the interior of the finitely ideal side of T and α^{i}_T the interior of the infinitely ideal side of T and α^{fi}_T the interior of the segment that is the intersection of the finitely ideal side and infinitely ideal side of T.

Incidentally, α^{fi}_T is an open line of $\mathbf{d}$-length π.

Theorems similar to the transversal intersection theorem of [**11**] hold for crescent-cones (and pseudo-crescent-cones). Let B be a convex i-ball in a convex j-ball C for $i < j$. Then B is *well-positioned* if B^o is a subset of C^o and δB that of δC. (It is fairly easy to list all types of well-positioned convex i-balls in a lune, a triangle, a tetrahedron, a trihedron or bihedron.)

The definition of the transversal intersection is quite difficult to write down but is fairly simple as Figure 10.1 shows.

DEFINITION 10.2. Let R_1 and R_2 be both crescent-cones. We say that R_1 and R_2 intersect transversally if the following statements hold ($i = 1, j = 2$, or $i = 2, j = 1$):

1. R_1 and R_2 overlap.
2. $\nu_{R_1} \cap \nu_{R_2}$ is a well-positioned radial segment s in ν_{R_1} and similarly in ν_{R_2}.
3. $\nu_{R_i} \cap R_j$ equals the closure of a component of $\nu_{R_i} - s$, and it is a well-positioned convex triangle D in R_j. Two sides of D other that s are finitely ideal and infinitely ideal respectively. (The three sides of D are well-positioned in three sides of R_j respectively.)
4. $R_i \cap R_j$ is the closure of a component of $R_i - D$.
5. Let t be the union of two sides of D other than s. Then t intersected with the infinitely ideal side of R_i is a well-positioned segment, and $\alpha^{\mathsf{i}}_{R_i} \cap \alpha^{\mathsf{i}}_{R_j}$ equals one of two components of $\alpha^{\mathsf{i}}_{R_i} - t$. Hence, $\alpha^{\mathsf{i}}_{R_i} \cup \alpha^{\mathsf{i}}_{R_j}$ is an open 2-disk. The same statement holds for the finitely ideal sides $\alpha^{\mathsf{f}}_{R_i}$ and $\alpha^{\mathsf{f}}_{R_j}$.
6. $t \cap \alpha^{\mathsf{fi}}_{R_i}$ is a point. $\alpha^{\mathsf{fi}}_{R_i} \cap \alpha^{\mathsf{fi}}_{R_j}$ equals one of two components of $\alpha^{\mathsf{fi}}_{R_i} - t$, and $\alpha^{\mathsf{fi}}_{R_i} \cup \alpha^{\mathsf{fi}}_{R_j}$ is an open line.

REMARK 10.1. We now give another characterization of transversal intersection, which is easily shown to be equivalent. R_1 and R_2 overlap, $\mathbf{dev}(\alpha^{\mathsf{i}}_{R_1})$ and $\mathbf{dev}(\alpha^{\mathsf{i}}_{R_2})$ are subsets of a common 2-hemisphere in $\mathbf{S}^2_\infty$ bounded by a great circle including both $\mathbf{dev}(\alpha^{\mathsf{fi}}_{R_1})$ and $\mathbf{dev}(\alpha^{\mathsf{fi}}_{R_2})$, and $\mathbf{dev}(\nu_{R_1})$ and $\mathbf{dev}(\nu_{R_2})$ meet transversally in a well-positioned segment.

EXAMPLE 10.1. From Example 6.1, we can easily obtain two crescent-cones and in general they meet transversally.

THEOREM 10.2. *Let R_1 and R_2 are overlapping crescent-cones. Then either $R_1 \subset R_2$ or $R_2 \subset R_1$ hold or R_1 and R_2 intersect transversally.*

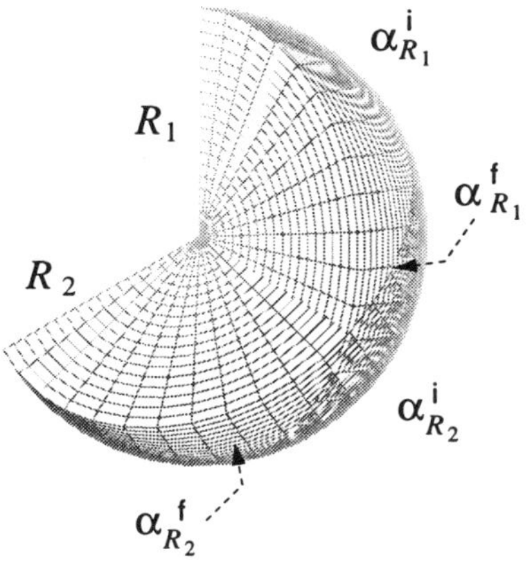

FIGURE 10.1. The transversal intersection of crescent-cones.

PROOF. The proof is similar to that of Theorem A.1. Only technical differences exist although we have to use the following lemma 10.1 instead of Lemma 5.2 of [**15**]. □

LEMMA 10.1. *Let N be a closed real projective manifold with a developing map* $\mathbf{dev} : \check{N}_h \to \mathbf{S}^3$ *for the completion $\check{N}_h$ of the holonomy cover N_h of N. Suppose that* $\mathbf{dev}$ *is an imbedding onto a union of two radiant bihedra H_1 and H_2 in the 3-hemisphere $\mathcal{H}$ meeting in a radiant bihedron or a radiant trihedron. Then $H_1 = H_2$, and N is convex.*

PROOF. The proof is similar to that of Lemma 5.2 in [**15**]. That is, we find a holonomy-group-invariant codimension-one submanifold in $H_1^o \cup H_2^o$ if H_1 is distinct from H_2. □

Let $\mathcal{R}_1$ be a collection of all crescent-cones in M_h. We define a relation that $R \sim S$ for $R, S \in \mathcal{R}_1$ if they overlap, i.e., R^o and S^o meet. (Recall that since R^o and S^o are open, this is equivalent to the statements that R and S^o meet or R^o and S meet.) Using this, we define an equivalence relation on $\mathcal{R}_1$, i.e., $R \sim S$ if there exists a finite sequence $R^1, R^2, \dots, R^n$ in $\mathcal{R}_1$ such that $R^1 = R$ and $R^n = S$ where R^i and R^{i+1} overlap for each $i = 1, \dots, n-1$.

We define the following sets as in [**11**] and exhibit their equivariant properties:

$$\Lambda^{\mathsf{c}}(R) = \bigcup_{S \sim R} S, \quad \delta^{\mathsf{i}}_\infty \Lambda^{\mathsf{c}}(R) = \bigcup_{S \sim R} \alpha^{\mathsf{i}}_S, \quad \delta^{\mathsf{f}}_\infty \Lambda^{\mathsf{c}}(R) = \bigcup_{S \sim R} \alpha^{\mathsf{f}}_S,$$
$$\delta^{\mathsf{fi}}_\infty \Lambda^{\mathsf{c}}(R) = \bigcup_{S \sim R} \alpha^{\mathsf{fi}}_S, \quad \Lambda^{\mathsf{c}}_1(R) = \bigcup_{S \sim R} (S - \nu_S) \quad \text{for } R \in \mathcal{R}_1 \tag{10.1}$$

DEFINITION 10.3. A *concave-cone affine manifold* N is an affine manifold such that N_h is a subset of $\Lambda^{\mathsf{c}}(R)$ for a crescent-cone R in $\check{N}_h$.

We list the properties of $\Lambda(R)$.

1. $\Lambda^{\mathsf{c}}(R)$ and $\Lambda^{\mathsf{c}}_1(R)$ are path-connected.

2. $\Lambda^c(R) \cap M_h$ is a closed radiant subset of M_h with totally geodesic boundary, which is also a radiant set.
3. $\Lambda^c_1(R)$ is a real projective 3-manifold with boundary $\delta_\infty \Lambda^c(R)$ defined as $\delta^{\mathsf{i}}_\infty \Lambda^c(R) \cup \delta^{\mathsf{f}}_\infty \Lambda^c(R) \cup \delta^{\mathsf{fi}}_\infty \Lambda^c(R)$ if given an obvious real projective structure induced from each crescent-cone. (Actually, the manifold-structure has corners [**32**]).
4. $\delta^{\mathsf{i}}_\infty \Lambda^c(R)$ is an open surface in $M^{\mathsf{i}}_{h\infty}$.
5. $\delta^{\mathsf{f}}_\infty \Lambda^c(R)$ is an open surface in $M^{\mathsf{f}}_{h\infty}$.
6. $\delta^{\mathsf{fi}}_\infty \Lambda^c(R)$ is an arc in $M^{\mathsf{i}}_{h\infty}$. The union $\delta_\infty \Lambda^c(R)$ is an open surface in $M_{h\infty}$ which maps totally geodesic except at $\delta^{\mathsf{fi}}_\infty \Lambda^c(R)$ under **dev**.
7. $\Lambda^c(R)$ is maximal, i.e., for any triangle T in $\check{M}_h$ with sides s_1, s_2, and s_3, if $s_2, s_3 \subset \Lambda^c(R) \cap M_h$, then $T \subset \Lambda^c(R)$.
8. $\mathrm{bd}\Lambda^c(R) \cap M_h$ is a properly imbedded countable union of surfaces in M^o_h, each component of which is a totally geodesic, radiant, properly imbedded open triangle or lune.
9. For any deck transformation ϑ, we have $\Lambda^c(\vartheta(R)) = \vartheta(\Lambda^c(R))$, $\delta^{\mathsf{i}}_\infty \Lambda^c(\vartheta(R)) = \vartheta(\delta^{\mathsf{i}}_\infty \Lambda^c(R))$, $\delta^{\mathsf{f}}_\infty \Lambda^c(\vartheta(R)) = \vartheta(\delta^{\mathsf{f}}_\infty \Lambda^c(R))$, and $\delta^{\mathsf{fi}}_\infty \Lambda^c(\vartheta(R)) = \vartheta(\delta^{\mathsf{fi}}_\infty \Lambda^c(R))$.

The proofs of the above items can be obtained by adding radial dimension to the proofs for real projective surfaces (see [**11**]). Since there is nothing surprising in this extension, we will not give proofs here.

EXAMPLE 10.2. In Example 6.1, we see that $\Lambda^c(R)$ for any crescent-cone R equals the closure of $U - l$, i.e., $\check{\mathcal{E}}_2$. $\delta^{\mathsf{f}}_\infty \Lambda^c(R)$ equals the xy-plane with O removed; $\delta^{\mathsf{i}}_\infty \Lambda^c(R)$ equals the interior of the hemisphere that is the intersection of $\mathbf{S}^2_\infty$ with the closure of U with the endpoints of l removed.

PROPOSITION 10.2. *The collection consisting of elements of form $\Lambda^c(R)$ for some crescent-cone R is locally finite, i.e., for any point x of M_h, there exists a*

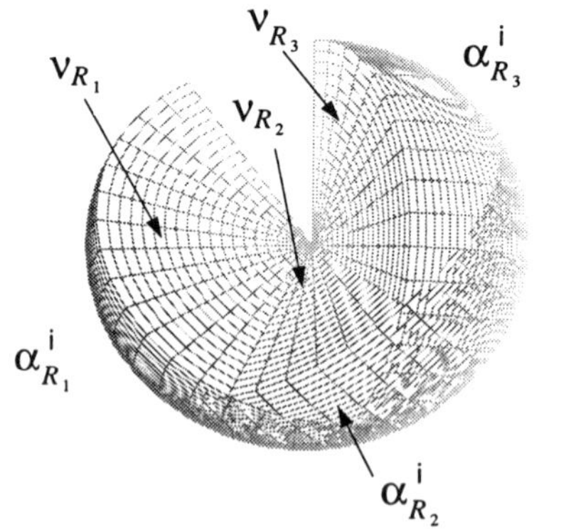

FIGURE 10.2. A picture of $\Lambda^c(R)$ when there are only three crescent-cones. Imagine when there are infinitely many. **dev** restricted to it might be something like the universal covering map of a solid torus, but the map is usually not a covering map.

neighborhood which intersects only finitely many elements of form $\Lambda^c(R)$. *(Note if two sets of form* $\Lambda^c(R)$ *are equal, we don't count twice.)*

PROOF. Suppose not. Let x be a point of M_h, and $B(x)$ a tiny ball of x and $B'(x)$ a closed tiny-ball-neighborhood of x in $B(x)^o$. Assume that our conclusion does not hold. Then there exists a sequence of crescent-cones R_i such that R_i meets $\text{int}B'(x)$ for each i and R_i is equivalent to a crescent-cone S_i where $\Lambda^c(S_i)$ are mutually distinct. We see from the definition of $\Lambda^c(R)$ that R_i are mutually inequivalent.

If $B'(x)$ is a subset of R_i for some i, then R_j and R_i overlap for each j, which is a contradiction. Hence, $\nu_{R_i} \cap B'(x)$ is not empty; let p_i be a point of $\nu_{R_i} \cap B'(x)$. We may assume without loss of generality that p_i converges to a point p of $B'(x)$ and the **d**-outer normal vector n_i to ν_{R_i} at p_i converges to a unit vector at p. Since $\alpha^{\mathsf{f}}_{R_i} \cup \alpha^{\mathsf{i}}_{R_i} \cup \alpha^{\mathsf{fi}}_{R_i}$ does not meet $B(x)$, Lemma B.2 shows that there exists a common open ball in $B(x)$ in R_i for i sufficiently large; that is, $R_i \sim R_j$ for i, j sufficiently large, a contradiction to the above paragraph. □

As an aside, we have:

PROPOSITION 10.3. *Suppose that a convex* 3*-ball* B *in* $\check{M}_h$ *is such that* δB *is a union of a convex* 2*-disk* ν_B *and a disk* α_B *in* δB *so that* ν_B *meets* M_h *and* α_B *is a subset of* $M_{h\infty}$. *Suppose that* B *satisfies the equivariance property, i.e., for each deck transformation either* $B = \vartheta(B)$ *or* $B \cap \vartheta(B) \cap M_h = \emptyset$. *Then the collection consisting of elements of form* $\vartheta(B)$ *for deck transformations* ϑ *is locally finite.*

PROOF. If the collection is not locally finite, then there exists a sequence $\vartheta_i(B)$ such that $\vartheta_i(B) \cap \vartheta_j(B) \cap M_h = \emptyset$ and $\vartheta_i(B)$ meets a tiny ball $B(x)$ of x for every i. We get contradiction similarly to above by Lemma B.2. □

PROPOSITION 10.4. *Given two crescent-cones* R *and* S, *we have one of three possibilities*:

- $\Lambda^c(R) = \Lambda^c(S)$ *and* $R \sim S$,
- $\Lambda^c(R)$ *meets* $\Lambda^c(S)$ *at the union of common components of* $\text{bd}\Lambda^c(R) \cap M_h$ *and* $\text{bd}\Lambda^c(S) \cap M_h$ *where* R *and* S *are not equivalent.*
- $\Lambda^c(R) \cap M_h$ *and* $\Lambda^c(S) \cap M_h$ *are disjoint.*

PROOF. The proof is same as in the results in Chapter 7 of [**15**]. The role played by n-crescent is played by crescent-cones. □

Suppose that $\text{bd}\Lambda^c(R) \cap M_h$ is empty. Then clearly $M_h \subset \Lambda^c(R)$ for a single crescent-cone R, and M is a concave-cone affine 3-manifold and we are done. We now assume that $\text{bd}\Lambda(R) \cap M_h$ is not empty for each crescent-cone R in $\check{M}_h$.

As in Chapter 7 of [**15**], a component $\text{bd}\Lambda^c(R) \cap M_h$ which is also a component of $\text{bd}\Lambda^c(S) \cap M_h$, for $R \not\sim S$, is said to be a *copied* component of $\text{bd}\Lambda^c(R) \cap M_h$. As in [**15**], we can show that a copied component is a common component of $\nu_T \cap M_h$ for some crescent cone T, $T \sim R$ and $\nu_U \cap M_h$ for U, $U \sim S$. Two copied components are either disjoint or identical, and the collection of all copied components are locally finite. (This is proved similarly to the arguments in Propositions 6.4 and 7.6 in [**15**].) The union of all copied components of $\text{bd}\Lambda^c(R) \cap M_h$ where R runs over all crescent-cones is denote by P_{M_h} and is said to be a *pre-two-faced surface.* It is a properly imbedded 2-manifold in M_h, and moreover by Proposition 2.1, $p|P_{M_h}$, for the covering map $p : M_h \to M$, covers a properly imbedded compact surface

A_M. Since each component of $\mathrm{bd}\Lambda^c(R)$ is a radiant set, each component of A_M is radially foliated. Hence A_M is a finite union of disjoint totally geodesic tori or Klein bottle. A_M is said to be the *two-faced submanifold.*

Note that the interior of every crescent-cone in $\check{M}_h$ is disjoint from P_{M_h}. Otherwise, a component of P_{M_h} and R^o for a crescent-cone R meets. Since each component of P_{M_h} is a components of $\nu_T \cap M_h$ for a crescent-cone T. We have $R \sim T$ and this means that R^o is a subset of $\mathrm{int}\Lambda^c(S)$ for $S \sim T$, a contradiction.

To see some examples of these, the Benzécri suspension of a real projective surface in Example 7.9 of [**15**] will give us a two-faced surface and sets of form $\Lambda(R)$. Further examples may be obtained from Benzécri suspensions of real projective surfaces.

As in [**15**], we can split M along A_M. Let N be the split manifold, and M' the split manifold of M_h by P_{M_h}. Then M' contains a copy of $M_h - A$ and $\mathbf{dev}|M_h$ extends to a projective map on M', again denoted by $\mathbf{dev}$. By pulling-back the metric μ and getting a path-metric $\mathbf{d}$ on M', we complete M' to a space $\check{M}'$.

PROPOSITION 10.5. *The crescent-cones in $\check{M}'$ and $\check{M}_h$ are in one-to-one correspondence given by $R \leftrightarrow R'$ if and only if $R^o = R'^o$ in $M_h - P_{M_h}$.*

PROOF. See Chapter 8 of [**15**]. □

We can easily show by Proposition 10.5 that M' contains no copied components as in [**15**] since we split M to obtain N. Hence, given any two crescent-cones R and S in $\check{M}'$, we have either $\Lambda^c(R) = \Lambda^c(S)$ or $\Lambda^c(R) \cap M'$ and $\Lambda^c(R) \cap M'$ are disjoint.

The holonomy cover N_h of N is obtained by taking a disjoint union of single appropriate components of M' for all components of N (see Chapter 8 of [**15**]). The developing map $\mathbf{dev}$ is obtained by restricting $\mathbf{dev} : M' \to \mathbf{S}^3$ to N_h, and the holonomy homomorphism is given by $\vartheta \mapsto h(\vartheta)$ when ϑ is a deck transformation of M' acting on a component of N_h. (Note here, we do not consider the action of deck transformations switching components.) $\mathbf{dev}|N_h$ is again a developing map, and we obtain the Kuiper completion $\check{N}_h$ of N_h using this developing map.

Obviously the split manifold N is still 2-convex. If not, $\check{N}_h$ includes a 3-crescent by Theorem 4.6 of [**15**] but $\check{N}_h$ cannot include radiant bihedra and hence 3-crescents by Propositions 10.1 and 3.6.

Let A be $\bigcup_{R \in \mathfrak{R}_1(N)} \Lambda^c(R) \cap N_h$ where $\mathfrak{R}_1(N)$ is the set of representatives of the equivalence classes in the set $\mathfrak{R}_1(N)$ of all crescent-cones in $\check{N}_h$. By Proposition 10.2, the collection of such subsets is locally finite, and $A \cap N_h$ is a submanifold of N_h by above paragraph since each $\Lambda^c(R') \cap N_h$ for a crescent-cone R' in $\check{N}_h$ is a submanifold. Therefore by Proposition 2.1, $P|A : A \to p(A)$ is a covering map onto a compact submanifold $p(A)$ in N of codimension 0 with nonempty boundary. Since each component of $\mathrm{bd}\Lambda^c(R) \cap N_h$ is totally geodesic (see the properties) and radiant, i.e., foliated by radial lines, each component of $p(A)$ is a compact submanifold with boundary whose components are totally geodesic tori or Klein bottles. Thus, we see that N decomposes into $p(A)$ and the closure of components of $N - p(A)$.

The holonomy cover of a component S of the closure of $N - p(A)$ can be considered a component S_h of the closure of $N_h - A$ with the induced covering map $p : S_h \to S$.

PROPOSITION 10.6. *Let A be a closed submanifold in $\check{N}_h$. Then the collection of crescent-cones in the Kuiper completion of $\check{A}$ of A are in one-to-one correspondence with those of $\check{N}_h$ in* $\mathrm{Cl}(A)$ *by $R \leftrightarrow R'$ if and only if $R^o = R'^o$.*

PROOF. See Section 8 of [**15**]. The proof can be obtained from the arguments there. □

Each component S of the closure of $N - p(A)$ is obviously 2-convex. Otherwise, $\check{S}_h$ must include a 3-crescent by Proposition 4.6 of [**15**] but this possibility was ruled out earlier by Proposition 10.1.

The components of the closure of $N - p(A)$ are convex since otherwise, the completion of the holonomy cover of at least one of them includes a crescent-cone by Theorem 6.1. Hence $\check{S}_h$ includes a crescent-cone for some component S, and by Proposition 10.6, $\mathrm{Cl}(S_h)$ in $\check{N}_h$ includes a crescent-cone in $\check{N}_h - A$, where S_h is regarded as a subset of N_h. Since for each crescent-cone R in $\check{N}_h$, $R \cap N_h$ is a subset of A by the definition of A, $R \cap N_h$ is a subset of the intersection of S_h and A. Since $R \cap N_h$ includes R^o, an open set and $S_h \cap A$ does not contain an open set, this is a contradiction. Thus each component S is convex.

Let K be a component of $p(A)$. Then K is covered by $\Lambda^c(R) \cap M_h$ for some crescent-cone R. By a reason similar to that in Chapter 8 of [**15**], i.e., Lemma 8.1, we can show that $\Lambda^c(R) \cap M_h$ is a holonomy cover of K with developing map $\mathbf{dev}|\Lambda^c(R) \cap M_h$. The surface $\Lambda^c(R) \cap M_h$ is a subset of $\Lambda^c(R')$ in the completion $(\Lambda^c(R) \cap M_h)\check{}$ of $\Lambda^c(R) \cap M_h$ for the crescent-cone R' in the completion with the identical interior as R by Proposition 10.6. Thus, each component of $p(A)$ is a concave-cone affine manifold with totally geodesic boundary, and this proves Theorem 10.1.

Theorems 11.1, 11.2, and 10.1 complete the proof of Theorem A. (Of course, assuming the result of Chapter 12.)

PROOF OF COROLLARY A. Suppose that M is closed. If M decomposes as in Theorem A, then since the decomposing surfaces are tangent to the radial flow, Theorems 3.3 shows that M admits a total cross-section.

If M does not decompose, then Theorem A shows that M is either a convex radiant affine 3-manifold or a generalized affine suspension as in (2) of Theorem A. By Theorem 3.3, M is a generalized affine suspension.

If M has nonempty boundary, and M decomposes nontrivially, then each boundary component of M is in one of the decomposed pieces. The decomposed manifolds are either convex, concave-affine or concave-cone affine. If a boundary component A is in a convex piece, then A is convex. If A is in the other two types, then a component A_h of the inverse image of A in the holonomy cover of the piece N_i, where A is, is a subset of a boundary of a crescent or a crescent-cone intersected with M_h. In these cases, A_h is affinely homeomorphic to a convex cone in $\mathbf{R}^2 - \{O\}$ and $\mathbf{R}^2 - \{O\}$ itself by Lemma 4.2. Theorem 3.4 shows that M is a generalized affine suspension.

Therefore, M admits a total cross-section to the radial flow by Proposition 3.2. The proof is completed by Theorem 3.1. □

PROOF OF COROLLARY B. Since a Benzécri suspension over an orbifold is always finitely covered by a Benzécri suspension over a surface obtained using the identity automorphism, and such a Benzécri suspension over a surface Σ is a trivial circle bundle over Σ (see Chapter 3), it follows that the Euler number is zero (see

Scott [**40**]). Thus, if M is a Benzécri suspension, then we are done. If M is a generalized affine suspension over a Euler characteristic nonnegative real projective surface, then M is homeomorphic to a bundle over a circle with fiber homeomorphic to the surface by considering only the topological aspects of the generalized affine suspension. If the Euler characteristic of a fiber is positive, then the fiber is either a sphere, $\mathbf{R}P^2$, or a disk, and M falls into the first case. If the Euler characteristic is zero, then M is as in the second case of the conclusion of the corollary. $\square$

CHAPTER 11

Concave-cone radiant affine 3-manifolds and radiant concave affine 3-manifolds

In this chapter, we will first classify concave-cone affine manifolds, which is an easy consequence of Tischler's work [**46**]. Next, we classify radiant concave affine manifolds. Let M be a compact radiant concave affine manifold with empty or totally geodesic boundary. Let M_h be its holonomy cover and $\check{M}_h$ the Kuiper completion. First, we do the classification for the case when there exists a unique radiant bihedron in $\check{M}_h$. We show that if there are three radiant bihedra in $\check{M}_h$ in general position, than M is finitely covered by a generalized affine suspension of a sphere. We will do this by showing that $\mathbf{dev}|\Lambda^{\mathsf{c}}(R)^o \cap M_h$ is a diffeomorphism onto the complement of a closed set K which is either a simply convex cone or the origin. If K is not the origin, then the deck-transformation group will act on the inverse image K' in M_h of $-K$ as an infinite cyclic group, leading to a desired contradiction. If K is the origin only, then M is shown to be a generalized affine suspension of a sphere or a real projective plane. Assume that there are no three radiant bihedra in $\check{M}_h$ in general position, which implies that the developing images of all radiant bihedra contain a common complete real line passing through O in their respective boundaries. We show that $\Lambda^o(R)$ equals M_h^o. If M has nonempty boundary, we easily obtain a contradiction by obtaining a deck-transformation-group invariant 3-crescent. If not, M_h is a cover of the complement of a complete real line in $\mathbf{R}^3$; such a radiant affine 3-manifold is shown not to exist by Barbot and Choi in Appendix C.

THEOREM 11.1. *Let M be a compact concave-cone affine 3-manifold with empty or totally geodesic boundary. Then M is a generalized affine suspension of affine tori, affine Klein bottles, affine annuli with geodesic boundary, or affine Möbius bands with geodesic boundary.*

Note that affine tori, affine Klein bottles, affine annuli with geodesic boundary, or affine Möbius bands with geodesic boundary are classified by Nagano-Yagi essentially (see Nagano-Yagi [**38**]). We use an argument communicated to us by T. Barbot.

PROOF. We have that $M_h \subset \Lambda^{\mathsf{c}}(R)$ for a crescent-cone R. Since $\delta^{\mathsf{f}}_\infty\Lambda^{\mathsf{c}}(R)$ is totally geodesic being a union of totally geodesic surfaces α^{f}_T, $T \sim R$, extending one another, $\mathbf{dev}(\delta^{\mathsf{f}}_\infty\Lambda^{\mathsf{c}}(R))$ is a subset of an affine plane P passing through O, and $\mathbf{dev}(\Lambda^{\mathsf{c}}(R))$ is a subset of the closure of a half-space H with boundary including P. Given any deck transformation ϑ of M_h, since $\vartheta(M_h) = M_h$, we have $\vartheta(\Lambda^{\mathsf{c}}(R)) = \Lambda^{\mathsf{c}}(R)$, and $\vartheta(\delta^{\mathsf{f}}_\infty\Lambda^{\mathsf{c}}(R)) = \delta^{\mathsf{f}}_\infty\Lambda^{\mathsf{c}}(R)$; thus, P is $h(\pi_1(M))$-invariant.

Introduce affine coordinates x, y, and z on $\mathbf{R}^3$ so that the origin O has coordinates $(0, 0, 0)$ and P is the xy-plane. Then the 1-form dz/z is invariant under

$h(\pi_1(M))$. This form induces a closed 1-form on M_h by **dev** which is invariant under the deck-transformation group, and the form descends to a closed 1-form transversal to the radial flow. Hence by Tischler's argument ([**46**]), it follows that the radial flow has a global section S (see Lemma C.3 of Appendix C).

There exists a closed hemisphere L which equals $\mathrm{Cl}(H) \cap \mathbf{S}^2_\infty$. Let L' be the interior of L. Let $\tilde{S}$ be a component of the inverse image of S in M_h. Then $\tilde{S}$ is transversal to radial flows and a subgroup G of the deck-transformation group acts on $\tilde{S}$ so that p induces a homeomorphism $\tilde{S}/G \to S$. For the radial projection Π from $\mathbf{R}^3 - \{O\}$ to $\mathbf{S}^2_\infty$, $\Pi \circ \mathbf{dev}|\tilde{S}$ maps into L' as an immersion. Since $h(\pi_1(M))$ acts on H, it acts on L'. By Berger [**8**], L' can be identified with an affine space and $h(\pi_1(M))$ acts as affine transformations on L'. Moreover, $\Pi \circ \mathbf{dev}|\tilde{S}$ is equivariant with respect to the homomorphism $i_{L'} \circ h$, where $i_{L'}$ is the restriction homomorphism to L'.

This means that S has an affine structure, and since $\delta S \subset \delta M$ and δM is empty or totally geodesic, it follows that δS is empty or geodesic. By the classification of affine surfaces, it follows that S is an affine torus, an affine Klein bottle, an affine annuli with geodesic boundary, or an affine Möbius band with geodesic boundary. Moreover, the existence of the total cross-section tells us that M is a generalized affine suspension over these affine surfaces by Proposition 3.2. □

THEOREM 11.2. *Let M be a compact radiant concave affine manifold with empty or totally geodesic boundary. Assume that M is not convex. Then M is a generalized affine suspension of a real projective hemisphere, a real projective plane, a real projective sphere, or a π-annulus (or Möbius band) of type C.*

Since M is radiant and concave affine, M_h is a subset of $\Lambda(R)$ for a radiant bihedron R in $\check{M}_h$. Recall that $\Lambda(R)$ is the union of all radiant bihedra equivalent to R. Given any radiant bihedron S in $\check{M}_h$, S^o meets a radiant bihedron equivalent to R, and hence $S \sim R$.

Suppose that there is only one radiant bihedron equivalent to R, namely itself. Then $M_h = R^o \cup (\nu_R^o \cap M_h)$. If M_h is closed, then $M_h = R^o$, which means $\nu_R^o \cap M_h = \emptyset$, a contradiction to the definition of 3-crescent. Thus, δM_h is not empty and equals $\nu_R^o \cap M_h$.

Since δM_h is not empty, M admits a total cross-section S to the radial flow by Theorem 3.4. Let $\tilde{S}$ be a component of $p^{-1}(S)$. Defining $\Pi : \mathbf{R}^3 - O \to \mathbf{S}^2_\infty$ be the radial flow, we see that $\Pi \circ \mathbf{dev}|\check{S}$ is an immersion into $\mathbf{dev}(\mathrm{Cl}(\alpha_R))$ equivariant with respect to $h|\pi_1(S)$. Furthermore, $\Pi \circ \mathbf{dev}|\tilde{S}^o$ is an imbedding onto the open hemisphere $\mathbf{dev}(\alpha_R)$. Thus, S is a real projective surface such that $\tilde{S}^o$ is real projectively isomorphic to an open hemisphere. By the following theorem 11.3, M is a generalized affine suspension of an hemisphere, a π-annulus or a π-Möbius band of type C.

THEOREM 11.3. *Let S be a compact real projective surface with non-empty geodesic boundary. Then if $\tilde{S}^o$ is real projectively diffeomorphic to an open hemisphere, then S is either a closed hemisphere, a π-annulus of type C or a π-Möbius band of type C.*

PROOF. Since $\tilde{S}^o$ is dense in $\tilde{S}$, its closure in $\check{S}$ equals $\check{S}$. Since $\tilde{S}^o$ is tame, $\check{S}$ is tame, and $\mathbf{dev} : \check{S} \to \mathbf{S}^2$ is an embedding onto its images $\mathbf{dev}(\check{S})$, a closed hemisphere H. Thus, we see that $\mathbf{dev}|\tilde{S}$ is an imbedding onto its image Ω, a domain whose interior is an open hemisphere H^o. Moreover, $\tilde{S}_\infty$ clearly maps into δH.

From the classification of real projective surfaces (see [**11**], [**12**], and [**13**]), if the Euler characteristic $\chi(S)$ of S is negative, either S itself is convex, or S^o contains a simple closed geodesic c such that its lift $\tilde{c}$ to $\tilde{S}^o$ is of **d**-length $< \pi$. The closure of the image of $\tilde{c}$ is a simply convex segment l with endpoints in $\tilde{S}_\infty$, and $\mathbf{dev}|l$ is a diffeomorphism to its image. However, there exists no simply convex line in H^o ending in δH in H.

If S is convex, then Lemma 1.5 of [**11**] shows that $\tilde{S}^o$ is simply convex. As H^o is not simply convex, this is a contradiction.

If $\chi(S) > 0$, then since S has nonempty boundary, S is diffeomorphic to a disk and hence S is a closed hemisphere.

If $\chi(S) = 0$, then assume that S is orientable, by taking a double cover if necessary. Since S is an annulus, $\tilde{S} \cap \delta H$ has two components on each of which the generator ϑ of $\pi_1(S)$ acts on. Since ϑ is orientation-preserving, we may assume that ϑ is of form (1)-(7) in Section 1.1 of [**11**] and the eigenvalues of ϑ is positive. by taking a double cover of S if necessary. Then since $\langle\vartheta\rangle$ acts without fixed points on H^o, ϑ is of type (2) or (4). (4) can be ruled out since one cannot get a compact quotient that is Hausdorff. Now, investigating when $\langle\vartheta\rangle$ acts properly and freely, we can easily show that S must be a π-annulus of type C. (We omit this somewhat tedious verification.) $\square$

We will now assume that there are more than one radiant bihedron. There has to be two overlapping bihedra since $\Lambda(R)$ contains $\check{M}$.

Let R_1, R_2, R_3 be radiant bihedra equivalent to R. We say that R_1, R_2, and R_3 are in general position if $\mathbf{dev}(\nu_{R_1})$, $\mathbf{dev}(\nu_{R_2})$, and $\mathbf{dev}(\nu_{R_3})$ are in general position.

PROPOSITION 11.1. *Suppose that there exists mutually overlapping three radiant bihedra R_1, R_2, and R_3 in general position. Then $\mathbf{dev}|\Lambda_1(R) \cap M_h$ is an imbedding onto $\mathbf{R}^3 - O$ or the complement of a closed simply convex radiant set K in $\mathbf{R}^3$.*

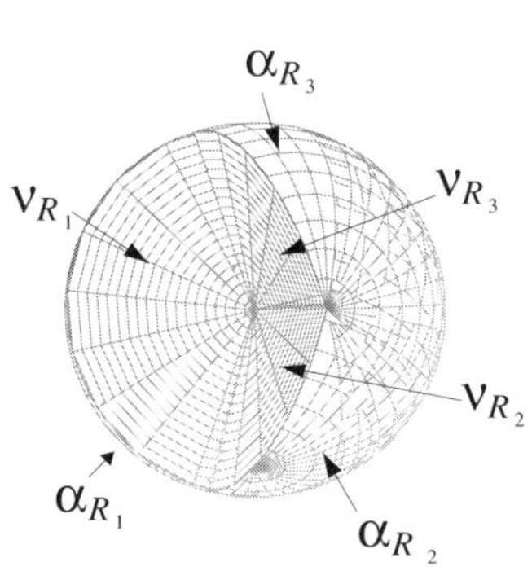

FIGURE 11.1. Three radiant bihedra in general position.

PROOF. First, $\mathbf{dev}|R_1 \cup R_2$ is an imbedding onto $\mathbf{dev}(R_1) \cup \mathbf{dev}(R_2)$ by Theorem A.1. Since $\mathbf{dev}(R_3) \cap (\mathbf{dev}(R_1) \cup \mathbf{dev}(R_2))$ is a connected 3-ball, it follows that $\mathbf{dev}|R_1 \cup R_2 \cup R_3$ is an imbedding onto $\mathbf{dev}(R_1) \cup \mathbf{dev}(R_2) \cup \mathbf{dev}(R_3)$ by Proposition 1.2.

We assume that all crescents denoted by different symbols are distinct in $\check{M}$ in this proof. Let S be an arbitrary radiant bihedron equivalent to R and overlapping some R_i for $i = 1, 2, 3$. Assume without loss of generality that S overlaps with R_3. $R_1 \cap R_3$ and $R_2 \cap R_3$ are two distinct radiant trihedra. From looking at images of these sets under the imbedding $\mathbf{dev}|R_3$, we see that $R_3 \cap S$ is a radiant trihedron and overlaps with at least one of $R_1 \cap R_3$ and $R_2 \cap R_3$. Assuming without loss of generality that $R_3 \cap S$ overlaps with $R_2 \cap R_3$, we see that S, R_2, and R_3 overlap mutually with one another. Hence $\mathbf{dev}|S \cup R_2 \cup R_3$ is an imbedding onto $\mathbf{dev}(S) \cup \mathbf{dev}(R_2) \cup \mathbf{dev}(R_3)$ as in the above paragraph.

Consider $\mathbf{dev}(R_1)^o \cap (\mathbf{dev}(R_2)^o \cup \mathbf{dev}(R_3)^o \cup \mathbf{dev}(S)^o)$. Since this also is a connected open 3-manifold by the following Lemma 11.1, Proposition 1.1 and Remark 1.3 show that $\mathbf{dev}|R_1^o \cup R_2^o \cup R_3^o \cup S^o$ is an imbedding. Since we may hence read the intersection pattern of R_1^o, R_2^o, R_3^o, and S^o by their image half-spaces in $\mathbf{R}^3$, we obtain that R_i, R_j, and S are mutually intersecting and are in general position for some choice of pair i, j, $i, j = 1, 2, 3$ (i.e., check that the other possibilities cannot occur).

A collection of radiant bihedra $S_1, \dots, S_n$ is said to be a *chain* if S_i overlaps with S_{i+1} for $i = 1, \dots, n-1$ with $n > 1$, or if $n = 1$. We say that a collection $R_1, R_2, R_3, S_1, \dots, S_n$ is a *chained* collection if any two elements of the collection are connected by a chain of overlapping elements in the collection. We will now prove by induction that given a chained collection $R_1, R_2, R_3, S_1, \dots, S_n$

(i) $\mathbf{dev}|R_1^o \cup R_2^o \cup R_3^o \cup S_1^o \cup S_2^o \cdots \cup S_n^o$ is an imbedding, and
(ii) given i, $i = 1, \dots, n$, S_i overlaps with R_j for some j, $j = 1, 2, 3$.

For $n = 1$, this was proved above. For $n = 2$, let R_1, R_2, R_3, S_1, S_2 be a chained collection. Assume without loss of generality that R_1, R_2, R_3, S_1 is a proper chained collection. By above, S_1 overlaps with R_i and R_j for some pair i, j, $i, j = 1, 2, 3$ and they are in general position. If S_2 overlaps with S_1, then an open trihedron $S_1^o \cap S_2^o$ must meet one of the open trihedra $S_1^o \cap R_i^o$ and $S_1^o \cap R_j^o$ by geometry and Proposition 1.2. Hence, we can assume that S_2 overlaps with at least one of R_1, R_2, and R_3, and (ii) follows.

Since the collection R_1, R_2, R_3, S_1 and the collection R_1, R_2, R_3, S_2 satisfy (i) and the intersection of the respective unions of the images of their interiors

$$(\mathbf{dev}(R_1)^o \cup \mathbf{dev}(R_2)^o \cup \mathbf{dev}(R_3)^o) \cup (\mathbf{dev}(S_1)^o \cap \mathbf{dev}(S_2)^o)$$

is clearly a connected open subset, (i) follows for the collection R_1, R_2, R_3, S_1, S_2 by Remark 1.3.

Assume that (i) and (ii) are true if $n \leq k$ for an integer k. By above, we assume that $k \geq 2$. We will now prove for $n = k + 1$. Let $R_1, R_2, R_3, S_1, \dots, S_{k+1}$ be a chained collection. We may assume without loss of generality that

$$R_1, R_2, R_3, S_1, \dots, S_m$$

is a proper chained subcollection with $m \leq k$ and S_{k+1} overlaps with S_l for $l \leq m$. Then by (ii) for $n \leq k$, $R_1, R_2, R_3, S_l, S_{k+1}$ is a chained collection, and hence $\mathbf{dev}|R_1^o \cup R_2^o \cup R_3^o \cup S_l^o \cup S_{k+1}^o$ is injective and S_{k+1} overlaps with R_j for some $j = 1, 2, 3$. (ii) follows for $n = k + 1$.

We claim that **dev** restricted to $R_1^o \cup R_2^o \cup R_3^o \cup S_1^o \cup \cdots \cup S_{k+1}^o$ is injective, which will prove (i) for $n = k+1$. If not, there exist points $p \in S_l$ and $q \in S_m$, some pair $l, m \leq k+1$, so that $\mathbf{dev}(p) = \mathbf{dev}(q)$. However since R_1, R_2, R_3, S_l and R_1, R_2, R_3, S_m are both chained collections by above, it follows that R_1, R_2, R_3, S_l, S_m is a chained collection, and **dev** restricted to $R_1^o \cup R_2^o \cup R_3^o \cup S_l^o \cup S_m^o$ is injective by (i) for $n = 2$, a contradiction.

From (i) for all k, it follows that $\mathbf{dev}|\Lambda_1(R) \cap M_h$ is injective. The complement of $\mathbf{dev}(\Lambda_1(R) \cap M_h)$ is disjoint from $\mathbf{dev}(R_1), \mathbf{dev}(R_2)$, and $\mathbf{dev}(R_2)$. Hence it must be simply convex. □

LEMMA 11.1. *Let L_1 be an open half space in $\mathbf{R}^3$, and L_2, L_3, L_4 open half spaces overlapping with L_1 whose respective boundary planes P_1, P_2, P_3, P_4 pass through O. If P_1, P_2, and P_3 are in general position, then $L_1 \cap (L_2 \cup L_3 \cup L_4)$ is a connected open 3-manifold.*

PROOF. We will show that $P_1 \cap (L_2 \cup L_3 \cup L_4)$ is a connected open set. Now our statement easily implies the lemma.

Since P_1, P_2, and P_3 are in general position, it follows that $P_1 \cap P_2$ and $P_1 \cap P_3$ are distinct lines passing through the origin. Thus, $P_1 \cap L_1$ and $P_1 \cap L_2$ are distinct half-planes meeting in an open set. Hence $P_1 \cap L_4$ must meet at least one of them in a nonempty open set, and $P_1 \cap (L_2 \cup L_3 \cup L_4)$ is a connected open set; this proves the above statement. □

Suppose that we are in the first case of the conclusion of Proposition 11.1. Then $\mathbf{dev}|\Lambda_1(R) \cap M_h$ is a homeomorphism onto $\mathbf{R}^3 - \{O\}$. Since $\Lambda_1(R) \cap M_h$ is dense in M_h, and $\mathbf{dev}(M_h)$ is a subset of $\mathbf{R}^3 - \{O\}$, it follows that $\Lambda_1(R) \cap M_h = M_h$, and $\mathbf{dev} : M_h \to \mathcal{H}^o - \{O\}$ is a homeomorphism, implying M is boundaryless. By the proof of Proposition 3.3, M is a generalized affine suspension over a real projective sphere or a real projective plane.

Suppose that we are in the second case; we will show this leads to contradictions. Since K is simply convex, the set $-K$ consisting of $-v$ for $v \in K$ is disjoint from $K - \{O\}$ and hence is a proper subset of $\mathbf{dev}(\Lambda_1(R) \cap M_h)$.

Since $\mathbf{dev}|\Lambda_1(R) \cap M_h$ is an imbedding onto the image by Proposition 11.1, it follows that there exists a set K' in $\Lambda^o(R)$ such that $\mathbf{dev}|K'$ is a diffeomorphism onto $-K - \{O\}$. For any deck transformation ϑ, since $\vartheta(M_h^o) = M_h^o$ and M_h^o is a subset of $\Lambda(R)$, it follows that by Proposition 10.4, $\vartheta(\Lambda_1(R)) = \Lambda_1(R)$. Hence $h(\vartheta)(\mathbf{dev}(\Lambda_1(R))) = \mathbf{dev}(\Lambda_1(R))$, K is $h(\pi_1(M))$-invariant, and the deck-transformation group G of M_h acts on K'.

Assume that K' has non-empty interior. Since the deck transformation group acts on K', the image $p(K')$ in M is a closed submanifold. Since K' is homeomorphic to $D^2 \times \mathbf{R}$ for a disk D^2 and is a radiant set, the boundary of K' which is homeomorphic to $\mathbf{S}^1 \times \mathbf{R}$ covers the boundary of K'/G, and hence the boundary of K'/G is homeomorphic to a torus or a Klein bottle. Since boundary is compressible, there exists an imbedded disk in K'/G with boundary $\delta K'/G$ by the loop theorem. Since K' is simply connected, K'/G is homeomorphic to a solid torus or a solid Klein bottle. Hence G is isomorphic to the group of integers $\mathbf{Z}$. If K' has empty interior, K' is homeomorphic to $I \times \mathbf{R}$ for an interval I or to $\mathbf{R}$. Since G acts on K' freely and properly, K'/G is homeomorphic to an annulus, a Möbius band, or a circle, and G is isomorphic to $\mathbf{Z}$.

Recall that the boundary of M is either empty or totally geodesic. Suppose the boundary of M is not empty. Let T be a component of δM, an affine torus or an affine Klein bottle. Since the fundamental group of $\pi_1(M)$ equals $\mathbf{Z}$, it follows that a simple closed curve α in T is a boundary of an imbedded disk D in M by the loop theorem. Since the holonomy of α is trivial, the classification of affine structures on tori or Klein bottles (see Nagano-Yagi [**38**]) shows that a component T' of $p^{-1}(T) \subset \delta M_h$ is affinely isomorphic to a finite cover of $\mathbf{R}^2 - \{O\}$.

Since δM_h is disjoint from $\Lambda_1(R) \cap M_h$, $\Lambda_1(R) \cap M_h$ is dense in M_h, and $\mathbf{dev}|\Lambda_1(R) \cap M_h$ is an imbedding onto $\mathbf{R}^3 - K$, it follows that $\mathbf{dev}(\delta M_h)$ is a subset of K. Since K is a simply convex subset, this contradicts the fact that an affine copy of a cover of $\mathbf{R}^2 - O$ contains complete affine lines. (A *complete affine line* is a complete affine one-dimensional subset of $\mathbf{R}^n$. It is real projectively homeomorphic to an open 1-hemisphere, and has $\mathbf{d}$-length π.)

Suppose next that δM is empty. Let t be a generator of $\pi_1(M)$. Then since any 3×3-matrix has a real eigenvalue, $h(t)$ must have a pair of antipodal fixed points in $\mathbf{S}^2_\infty$ (by taking a double cover of M if necessary). They correspond to a pair of antipodal radial line in $\mathcal{H}^o$. At least one of them, say l, is in $\mathbf{dev}(\Lambda_1(R) \cap M_h)$ since $\mathbf{dev}(\Lambda_1(R) \cap M_h)$ is the complement of a simply convex cone in $\mathbf{R}^3$. The radial line l' in M_h corresponding to l maps to a periodic orbit c in M. Suppose that c is of saddle type in Barbot's terminology (see Theorem 3.3). Then $h(t)$ has three distinct eigenvalues, i.e., diagonalizable. We may assume without loss of generality that $h(t)$ has positive eigenvalues (by taking a double cover of M if necessary). Then choosing the fixed point in $\mathbf{S}^2_\infty$ of attracting or repelling type, we obtain another radial line m in $\Lambda_1(R) \cap M_h$ fixed by t. Then m corresponds in M to a periodic orbit which is not of saddle type, and by Theorem 3.3, M is a generalized affine suspension of a projective surface Σ. If c is not of saddle type to begin with, then M is a generalized affine suspension of a projective surface Σ.

Since the fundamental group of M equals $\mathbf{Z}$, it follows that the projective surface Σ is a disk or sphere. The surface cannot be a disk since M is closed. If the surface is a sphere, then since the projective structure on the sphere is unique up to isotopy, it follows that M is obtained from a generalized affine suspension of the real projective sphere. Hence, our original M is covered by the generalized affine suspension of the real projective sphere. However this means that K is the origin only, a contradiction.

Now we can assume the following statement: Whenever R_1, R_2, and R_3 are mutually overlapping radiant bihedra, $\mathbf{dev}(\nu_{R_1}), \mathbf{dev}(\nu_{R_2})$, and $\mathbf{dev}(\nu_{R_3})$ include a common complete real line. We will show that such a case is impossible, and this will complete the proof of Theorem 11.2.

By induction, we will show that $\mathbf{dev}(\nu_S)$ for any $S \sim R$ includes l for a fixed complete affine line l. Let S be a radiant bihedron so that $S \sim R$. We choose a sequence $R_1, R_2, R_3, \dots, R_n$ so that R_i and R_{i+1} overlap for $i = 1, \dots, n-1$ and $R_n = S$, and $\mathbf{dev}(\nu_{R_i})$ contains a complete real line l for $i = 1, 2$.

We may assume that R_i are distinct from R_{i+1} for each i so that $\mathbf{dev}(\nu_{R_i})$ and $\mathbf{dev}(\nu_{R_{i+1}})$ are transversal. Assume that $\mathbf{dev}(\nu_{R_i})$ contains l for $i = 1, 2, \dots, k$ for $2 \leq k < n$. We will prove this for $k+1$. If $\mathbf{dev}(\nu_{R_{k+1}})$ does not contain l, then $\mathbf{dev}(\nu_{R_{k-1}})$, $\mathbf{dev}(\nu_{R_k})$, and $\mathbf{dev}(\nu_{R_{k+1}})$ are in general position. Since R_{k-1} and R_k overlap and R_k and R_{k+1} overlap, $\mathbf{dev}|R_{k-1} \cup R_k \cup R_{k+1}$ is an imbedding onto $\mathbf{dev}(R_{k-1}) \cup \mathbf{dev}(R_k) \cup \mathbf{dev}(R_{k+1})$ by Proposition 1.2. This means that R_{k-1}, R_k, and R_{k+1} are mutually overlapping radiant bihedra in general position. This case was ruled out above. Therefore, we conclude that $\mathbf{dev}(\nu_S)$ includes l for any $S \sim R$.

We will now prove that $M_h = \Lambda(R) \cap M_h$.

LEMMA 11.2. *$\mathbf{dev}(M_h)$ misses l.*

PROOF. Suppose that $\mathbf{dev}(x)$ belongs to l for $x \in M_h$. Since $M_h \subset \Lambda(R)$, there exists a radiant bihedron S containing x with $S \sim R$. Since there is more than one bihedral crescent in $\check{M}_h$, it follows that there is a bihedral crescent S_1 transversally meeting with S such that $\mathbf{dev}|S \cup S_1$ is an imbedding onto $\mathbf{dev}(S) \cup \mathbf{dev}(S_1)$, and $\mathbf{dev}(S)$ is distinct from $\mathbf{dev}(S_1)$.

Since $\mathbf{dev}(x) \in l$, we have $\mathbf{dev}(x) \in \mathbf{dev}(\nu_{S_1})$, and this means $x \in \nu_{S_1}$. Similarly, $x \in \nu_S$.

Note that if two properly imbedded surfaces meet at a point tangentially, then they must agree completely. If $x \in \delta M_h$, then since $x \in \nu_S$ for $S \sim R$, it follows that a component of $\nu_S \cap M_h$ equals a component of δM_h. Thus, if for another crescent S', $x \in \nu_{S'}$, then $\nu_S \cap M_h$ and $\nu_{S'} \cap M_h$ agree on this component. By Corollary 3.1, it follows that that $S = S'$ actually; however, our choice of S_1 contradicts this.

Suppose that $x \in M_h^o$. Let B be a small tiny ball neighborhood of x in M_h^o. Then $\mathbf{dev}|\bigcup_{t\in\mathbf{R}} \Phi_{h,t}(B)$ is an imbedding onto a simply convex cone C' since $\mathbf{dev}(x) \neq O$. Letting C equal to this union, we see again that $\mathbf{dev}|C \cup S^o \cup S_1^o$ is an imbedding onto $\mathbf{dev}(C) \cup \mathbf{dev}(S^o) \cup \mathbf{dev}(S_1^o)$. By the geometry of the situation, there exists a radiant bihedron S' in the image so that its sides, $\mathbf{dev}(\nu_S)$, and $\mathbf{dev}(\nu_{S_1})$ meet in general position. Let S'' be the inverse image in $C \cup S^o \cup S_1^o$. Then the closure S_2 of S'' in $\check{M}_h$ is a radiant bihedron such that S, S_1, and S_2 overlap mutually and $\mathbf{dev}(\nu_S), \mathbf{dev}(\nu_{S_1})$, and $\mathbf{dev}(\nu_{S_2})$ are in general position, contradicting what we just assumed. □

LEMMA 11.3. *$M_h = M_h^o = \Lambda_1(R) \cap M_h$, and hence $M_h^o \subset \Lambda_1(R)$.*

PROOF. Since S^o for any crescent S, $S \sim R$ is a subset of M_h^o, we obviously have $\Lambda_1(R) \cap M_h \subset M_h^o$.

We prove that $M_h^o \subset \Lambda_1(R)$. Suppose not. Then there exists a point x of M_h^o in $\mathrm{bd}\Lambda_1(R) \cap M_h$, which is not in T^o for any $T \sim R$. Since $\Lambda(R) \cap M_h$ is closed, we have $x \in \nu_S$ for some $S \sim R$ and by the above lemma, $\mathbf{dev}(x)$ is not in l.

Let l' denote the inverse image of l in ν_S. There has to be at least one radiant bihedron T overlapping with S and distinct from S. By Corollary 3.1, S and T intersect transversally. Since $\mathbf{dev}(\nu_S) \cap \mathbf{dev}(\nu_T)$ contains l, it follows that $\nu_S \cap \nu_T$ contains l', and there exists a component E of $\nu_S^o - l'$ contained in T^o as we can see from the conditions for transversal intersections in Appendix A. While E does not contain x since x does not belong to T^o, we conclude that a component, namely E, of $\nu_S^o - l'$ not containing x is an open lune completely contained in M_h^o.

Assume without loss of generality that $h(\pi_1(M))$ fixes p and $-p$ (by taking a double cover if necessary); that is, $h(\pi_1(M))$ acts on l preserving the orientation of $\mathbf{S}^3$. We claim that S cannot overlap with its image under a deck transformation distinct from S. If our claim is false, then S and $\vartheta(S)$ for a deck transformation ϑ intersect transversally by Corollary 3.1. By an orientation condition, x belongs to $\vartheta(S^o)$ or $\vartheta(x)$ belongs to S^o. In the latter case, x belongs to $\vartheta^{-1}(S^o)$. Both of these are impossible. Therefore, for any deck transformation ϑ of M_h, we have either $S = \vartheta(S)$, $S \cap \vartheta(S) \cap M_h = \emptyset$, or $S \cap \vartheta(S) \cap M_h$ is a union of common components of $\nu_S \cap M_h$ and $\nu_{\vartheta(S)} \cap M_h$ where S and $\vartheta(S)$ do not overlap as in [**15**].

For a component A of $\vartheta(\nu_S) \cap M_h$ for a deck transformation ϑ of M_h, we say that A is a *copied component for* $\vartheta(S)$ if A is a component of $\varphi(\nu_S) \cap M_h$ for a deck transformation φ where $\vartheta(S)$ and $\varphi(S)$ do not overlap. The union $\hat{A}$ of all copied components of $\vartheta(S)$ as ϑ ranges over deck transformations ϑ of M_h is a properly imbedded submanifold in M_h. The proof of this fact is similar to the proofs for pre-two-faced submanifolds for 3-crescents and crescent-cones above (see also [**15**]). Therefore, we see that $p(\hat{A})$ must be a properly imbedded closed surface in M.

We split off along $p(\hat{A})$ to obtain the result N. As before the holonomy cover N_h of N is a disjoint union of components of M_h split along $\hat{A}$. The Kuiper completion $\check{N}_h$ of N_h is obtained by completing with respect to the metric induced by $\mathbf{dev} : N_h \to \mathbf{S}^3$ which is the immersion extended from $M_h - A'$.

We can choose N_h so that N_h includes S^o. The closure S' of S^o in $\check{N}_h$ is a radiant bihedron in $\check{N}_h$. As in crescent cases [**15**], the splitting process insures that there is no copied component for $\vartheta(S')$ in $\check{N}_h$ where ϑ is a deck transformation of $\check{N}_h$. Lemma 11.4 shows that E must be a triangle, which is a contradiction.

If there are no copied components in $\nu_S \cap M_h$, then S satisfies the premise of Lemma 11.4 showing that E is a triangle, a contradiction. This shows that M_h^o is a subset of $\Lambda_1(R)$.

If δM_h is not empty, then a point x of δM_h is in $\mathrm{bd}\Lambda_1(R)$. x must be a point of a radiant bihedron S, $S \sim R$, which has the same property as the radiant bihedron denoted by S above. The same argument as above shows that this cannot happen. Hence it follows that $M_h = M_h^o$. □

LEMMA 11.4. *Suppose that given a radiant bihedron S, either we have $S = \vartheta(S)$ or $S \cap \vartheta(S) \cap M_h = \emptyset$ for any deck transformation ϑ. Suppose that a complete affine line l' in $M_{h\infty}$ lies in ν_S. Then $\nu_S \cap M_h$ is a union of two radiant convex triangles respectively in the two components of $\nu_S - l'$.*

PROOF. Since the collection whose elements are of form $\vartheta(S)$ is locally finite by Proposition 10.3, Proposition 2.1 implies that $S^o \cup (\nu_S \cap M_h)$ covers a compact three-dimensional submanifold K in N with boundary $p(\nu_S \cap M_h)$. Since K is a radiant affine 3-manifold it follows that K admits a total cross-section S by Theorem 3.4. Then by Theorem 11.3, K is a generalized affine suspension over a hemisphere, or a π-annulus (or Möbius band) of type C. The first case is not possible as $\nu_S \cap M_h$ has at least two components respectively in each side of $\nu_S - l'$. The second case implies the conclusion of the lemma. □

Given a lune L in $\mathcal{H}$ with boundary equal to l union a segment in $\mathbf{S}^2_\infty$ with endpoints p and $-p$, we say that $L - l$ is a *proper lune.* A subset J of a lune in $\check{M}_h$ such that $\mathbf{dev}(J)$ is a proper lune in $\mathcal{H}$ is said to be a *proper lune.*

LEMMA 11.5. $\mathbf{dev}|\Lambda_1(R) : \Lambda_1(R) \to \mathcal{H} - l$ *lifts to an imbedding* $\mathbf{dev}'$ *to some cover W of $\mathcal{H} - l$, and the image* $\mathbf{dev}'(M_h)$ *is a union of open radiant bihedra in W.*

PROOF. Let $\mathbf{S}^1$ be the great circle in $\mathbf{S}^2_\infty$ which is perpendicular to a segment on $\mathbf{S}^2_\infty$ connecting p and $-p$. Given a radiant bihedron S, $S \sim R$, let l'_S denote the inverse image of $\mathbf{dev}^{-1}(l)$ in S. Since for each point x of $\Lambda_1(R)$, $x \in T^o \cup \alpha_T$ for some $T \sim R$, it follows that x belongs to a unique proper lune J.

Let $\mathcal{P}$ denote the set of all proper lunes in $\Lambda_1(R)$. Given the topology of geometric convergence to $\mathcal{P}$ with respect to the metric $\mathbf{d}$, $\mathcal{P}$ becomes a differentiable 1-manifold (without boundary).

There exists an immersion $f : \mathcal{P} \to \mathbf{S}^1$ given by mapping J to the intersection of $\mathbf{S}^1$ with the closure of J. If $\mathcal{P}$ is simply connected, i.e., $\mathcal{P}$ is an open arc, then we let W be the infinite cyclic cover of $\mathcal{H} - l$. If $\mathcal{P}$ is homeomorphic to a circle, then let W be the n-fold cyclic-cover of $\mathcal{H} - l$ where n is the degree of f. Then $\mathbf{dev}|\Lambda_1(R)$ lifts to an injective map into W since no two proper lunes map into same one in W. Since $\mathbf{dev}|\Lambda_1(R)$ is an open map, the first statement follows. Since $M_h = \Lambda_1(R) \cap M_h$, $\mathbf{dev}'(M_h)$ is a union of open radiant bihedra in W. □

We let $q : W \to \mathcal{H} - l$ denote the covering map. We denote by W_∞ the part of W covering $\mathbf{S}^2_\infty - l$. Using the radiant vector field on W induced from that of $\mathcal{H} - l$, we define $\Pi : \mathcal{H} - l \to \mathbf{S}^2_\infty - l$ the radial projection and $\Pi_W : W \to W_\infty$ that induced from it so that $\Pi \circ \mathbf{dev}|M_h = q \circ \Pi_W \circ \mathbf{dev}'|M_h$.

We see that W_∞ is foliated by lines corresponding to lines in $\mathbf{S}^2_\infty$ with endpoints p and $-p$. Since $M_h = \Lambda_1(R) \cap M_h$, $\Pi_W \circ \mathbf{dev}'(M_h)$ is a union of the lines in this foliation $\mathcal{F}$. Letting $\mathbf{S}^1_W$ denote the inverse image of $\mathbf{S}^1$ in W_∞, we see that $\Pi_W \circ \mathbf{dev}'(M_h) \cap \mathbf{S}_W$ is a connected open set U.

Note that $\mathbf{S}^1_W$ has a path-metric $\mathbf{d}$ induced from the Riemannian metric pulled from $\mathbf{S}^1$ by the covering map. The $\mathbf{d}$-length of $\mathbf{S}^1_W$ is greater than equal to π since M_h includes at least one radiant bihedron. Since there are more than one radiant bihedra, we have $\mathbf{d}$-length $> \pi$.

Suppose that U has an endpoint q in $\mathbf{S}^1_W$. There exists a point q' which is at the $\mathbf{d}$-distance π away from q in U. Since $\mathbf{dev}'(M_h)$ is the union of open hemispheres $\mathbf{dev}'(\alpha_S)$ for some $S \sim R$, it follows that there exists an open hemisphere H in W_∞ such that $H \cap \mathbf{S}^1_W$ equals the arc connecting q and q', and H equals $\Pi_W(T^o)$ for a radiant bihedron T in $\mathbf{dev}'(\check{M}_h)$, where $T \sim R$. (To see this use the fact that $\mathbf{dev}'$ is an imbedding: $M_h^o \to W$.)

Since T is a lune such that $\Pi_W(T^o)$ contains an end of U, and we assumed that $h(\pi_1(M))$ acts on p and $-p$ and we assume that $h(\pi_1(M))$ acts on the foliation $\mathcal{F}$ in an orientation preserving manner (by taking a double cover of M if necessary), then the deck transformation group of M_h acts on T. At least one component of $\nu_T^o - l'_T$ is an open lune in M_h as in the proof of Lemma 11.3 since T has to overlap with at least one radiant bihedron different from itself. By Lemma 11.4, we get a contradiction as before. Therefore, $\mathbf{dev}'(M_h) \cap \mathbf{S}^1_W$ is infinitely long in both directions; or equals $\mathbf{S}^1_W$ itself and W is a cyclic cover of $\mathcal{H} - l$.

This means that $\mathbf{dev}'|\delta_\infty\Lambda(R)$ is a map onto W_∞, and $\mathbf{dev}'|M_h$ is an imbedding onto W. This is a contradiction by the following lemma.

LEMMA 11.6 (Barbot-Choi (Appendix C)). *A radiant affine 3-manifold N does not have the developing map $\mathbf{dev} : N_h \to \mathcal{H}^o$ that is a finite or infinite cyclic covering map onto $\mathcal{H}^o - l$.*

Obviously such N must be a closed manifold.

CHAPTER 12

The nonexistence of pseudo-crescent-cones

As promised in Chapters 6 and 10, we will show that pseudo-crescent-cones do not exist assuming that M is not convex. A radiant affine 3-torus, as constructed by J. Smillie [**42**], which is obtained as a quotient of a radiant tetrahedron in $\mathcal{H}$, the Kuiper completion, the closure of the tetrahedron, of the holonomy cover contains a pseudo-crescent-cone.

In this chapter, we will work on the Kuiper completion $\check{M}$ of $\tilde{M}$ as we can lift any pseudo-crescent on $\check{M}_h$ to $\check{M}$ by Lemma 12.1. We will denote by **dev** the developing map $\check{M} \to \mathbf{S}^3$ and by $\mathbf{d}$ the metrics on $\check{M}$ and $\check{M}_h$ induced from $\mathbf{S}^3$ by developing maps. (This is a slight abuse of notation.)

First, we define the notion of infinite pseudo-crescent cones generalizing that of pseudo-crescent cones for convenience of proof (see Definition 12.1). We will show that a subset of the boundary of a pseudo-crescent-cone or an infinite pseudo-crescent-cone cannot cover a compact submanifold in M (see Proposition 12.1) and some lemmas.

We now assume that there exists a pseudo-crescent-cone or an infinite pseudo-crescent cone in $\check{M}$ and try to obtain contradictions. We discuss the transversal intersections of two pseudo-crescent-cones. We show that infinite pseudo-crescent-cones are fictitious as M is assumed to be not convex. This follows from the equivariance property of the maximal infinite pseudo-crescent-cones. Next, we define the union of collections of pseudo-crescent-cones and show that the union satisfies nice properties such as equivariance, and $\tilde{M}$ is a subset of one of such unions. Next, we show that the union has triangles at the ideal boundary which can be linearly ordered. Since the deck-transformation group acts on the set of linearly ordered triangles, if the action is trivial, then $\pi_1(M)$ acts on a pseudo-crescent-cone R and hence a component of $\nu_R \cap \tilde{M}$ covers a closed surface, a contradiction by Proposition 12.1. If the action is nontrivial, it follows by 3-manifold topology that M is a fiber bundle over a circle with fiber homeomorphic to a closed surface, and, again, we obtain a contradiction by Proposition 12.1.

The sides of a compact radiant convex 3-ball P consist of radial segments, radiant triangles, and the unique side I_K in $\mathbf{S}^2_\infty$ so that P is the cone over I_K. A compact radiant convex 3-ball P in $\mathcal{H}$ is said to be an *infinite* polyhedron if the collection of its sides not in $\mathbf{S}^2_\infty$ is locally finite except at points of two radial edges of one side F of P and the collection of the sides are countably infinite. F is called the *fundamental side*, and an edge where the ideal sides are not locally finite a *special* edge of the fundamental side.

Again a convex compact ball and its parts in $\check{M}$ are named by how their images are named in $\mathbf{S}^3$.

DEFINITION 12.1. A radiant convex polyhedron T in $\check{M}$ that has only a side meeting $\tilde{M}$ and the rest lying in $\check{M}_\infty$ is said to be a *pseudo-crescent-cone*. We

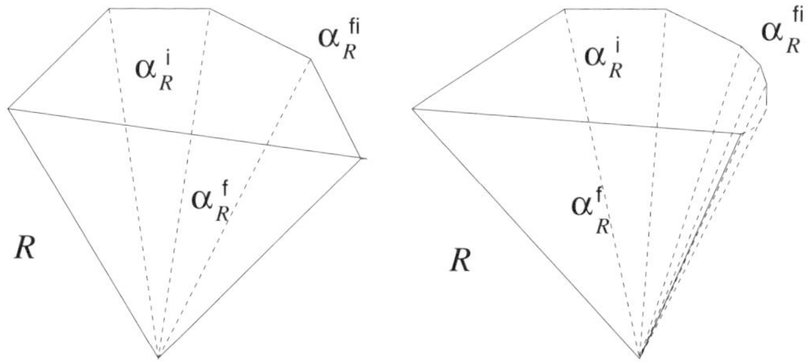

FIGURE 12.1. The pictures of a pseudo-crescent-cone and an infinite pseudo-crescent-cone.

define ν_T be the side that meet $\tilde{M}$ and α_T^{i} the interior of the infinitely ideal side of T, α_T^{f} the interior of the disk that is the union of the finitely ideal sides of T, and α_T^{fi} the interior of the arc that is the union of all segments in the intersection of the infinitely ideal side and the union of the finitely ideal sides.

DEFINITION 12.2. A radiant convex infinite polyhedron T in $\check{M}$ with the fundamental side meeting $\tilde{M}$ and the rest of the sides in $\tilde{M}_\infty$ is said to be an *infinite pseudo-crescent-cones*. We define ν_T be the fundamental side and α_T^{i} the interior of the infinitely ideal side of T and α_T^{f} the interior of the disk that is the union of the finitely ideal sides of T, and α_T^{fi} the open arc that is the interior of the arc that is the intersection of the infinitely ideal side and the union of the finitely ideal sides.

Pseudo-crescent-cones and infinite pseudo-crescent-cones on $\check{M}_h$ are defined the same way as for $\check{M}$, and various parts of them are also defined as above.

LEMMA 12.1. *If $\check{M}_h$ includes a pseudo-crescent-cone (resp. infinite pseudo-crescent-cone), then $\check{M}$ includes a pseudo-crescent cone (resp. infinite pseudo-crescent-cone).*

PROOF. Let R be a pseudo-crescent-cone in $\check{M}_h$. Lift the simply convex open 3-ball R^o to $\check{M}$, and denote the image of the lift by P. Then since P is a simply convex open 3-ball also, $\mathbf{dev}|P$ is an imbedding onto a simply convex open 3-ball $\mathbf{dev}(P) = \mathbf{dev}(R^o)$, and the closure $\mathrm{Cl}(P)$ of P in $\check{M}$ is a simply convex 3-ball. Since the covering map $c : \tilde{M} \to M_h$ is distance non-increasing with respect to $\mathbf{d}$, it extends to a map $\check{c} : \check{M} \to \check{M}_h$. Since $c|P : P \to R^o$ is a $\mathbf{d}$-isometry, $\check{c} : \mathrm{Cl}(P) \to R$ is an embedding.

We note that under $\check{c}$, $\tilde{M}$ maps into M_h obviously. Hence $\check{c}^{-1}(M_{h\infty}) \subset \tilde{M}_\infty$. We also have $\check{c}^{-1}(M_h) \subset \tilde{M}$: Otherwise a point p of $\tilde{M}_\infty$ maps to a point p' in M_h. We may find a point q in $\tilde{M}$ and a path α in $\check{M}$ connecting q to p. Then $c \circ \alpha$ is a path in M_h with endpoint p' and q' which lifts to α. Since c is a covering map, this means $p \in \tilde{M}$.

From above, it follows that the inverse image of $R \cap M_{h\infty}$ under $\check{c}|\mathrm{Cl}(P)$ is a subset of $\tilde{M}_\infty$ and that of $R \cap M_h$ is a subset of $\tilde{M}$; that is, $\mathrm{Cl}(P)$ is a pseudo-crescent. When R is an infinite pseudo-crescent cone, a similar argument applies. □

The converse of the above lemma is also true, but we won't prove it.

PROPOSITION 12.1. *Let F be a component of $\nu_R \cap \tilde{M}$ for a pseudo-crescent-cone or infinite pseudo-crescent-cone R. Then $p|F$ is not a covering map of an imbedded closed surface in M.*

PROOF. The component F is an open triangular component of $\nu_R \cap \tilde{M}$. If F covers a closed surface S in M, then there exists a subgroup G of deck transformations acting on F so that $p|F : F \to S$ induces a homeomorphism $F/G \to S$. Up to choosing an index two subgroup, we may assume that G preserves the sides of the submanifold F, and G acts on R by Propositions 12.2 and 12.3 and acts on $\nu_R \cap \tilde{M}$.

Since F is a radiant open set, F is a simply connected disk. By Lemma 12.2, F is not the only component of $\nu_R \cap \tilde{M}$. Since G has to act on F and ν_R, and F is a proper cone in ν_R, it follows that $h(G)$ has at least three fixed points on the top side of $\mathbf{dev}(\nu_R)$. Thus for every element g of G, $h(g)$ restricts to a dilatation, i.e., of form $s\mathrm{I}$, $s \in \mathbf{R}^+$, in the hyperplane containing $\mathbf{dev}(\nu_R)$, and so $\mathbf{dev}(F)/h(G)$ is not compact. Since the imbedding $\mathbf{dev}|F : F \to \mathbf{dev}(F)$ induces a homeomorphism $F/G \to \mathbf{dev}(F)/h(G)$, this is a contradiction. □

LEMMA 12.2. *Suppose that N is a real projective n-manifold and $\tilde{N}$ includes a convex n-ball B such that $B^o \subset \tilde{N}$ and $\mathrm{bd}B \cap \tilde{N}$ is an open (connected) convex subset F of a side of B. Suppose that F covers a compact $(n-1)$-dimensional submanifold in N. Then N is convex.*

PROOF. By lifting to a finite cover, we assume that $p(F)$ and N are orientable.

Since F is a cover of an imbedded closed submanifold F/G in N, for any element ϑ of $\pi_1(N)$, we have either $\vartheta(F) = F$ or $\vartheta(F)$ and F are disjoint, and, moreover, the collection $\{\vartheta(F)|\vartheta \in \pi_1(N)\}$ is locally finite in $\tilde{N}$.

If F is a subset of $\delta\tilde{N}$, then clearly B^o includes $\tilde{N}^o$. Since $\delta B \cap \tilde{M} = F$, it follows that $\tilde{N} = B^o \cup F$, which means that $\tilde{N}$ is convex.

Assume now that F is a subset of $\tilde{N}^o$. Then F separates $\tilde{N}$ into two parts, one of which is B^o and the other is $\tilde{N} - B$.

If $N - p(F)$ is connected, then there exists a simple closed curve α in N intersecting with $p(F)$ only once. This implies that the homotopy class of α is not trivial, and that there are infinitely many copies of F in B^o obtained by applying the elements of $\pi_1(N)$ that are powers of the deck transformation corresponding to α.

Suppose that $N - p(F)$ has two components, and there are only finitely many copies of F in B under the deck-transformation-group action. Then we choose a copy F' so that a component of $B - F'$ not including F includes no copies of F. Let B', $B' = B'^o \cup F$, denote the closure of this component. Then F' separates $\tilde{N}$ into a convex set B'^o and $\tilde{N} - B'$. Since B'^o is a component of $\tilde{N} - p^{-1}(p(F))$, we have that B'^o must cover a component of $N - p(F)$, and hence, $B'^o \cup F$ covers the closure N_1 of a component of $N - p(F)$. Each deck transformation acting on B'^o acts on F nontrivially, and a deck transformation acting on F acts on B'^o since N and $p(F)$ are orientable and F is the only boundary component of B', seen as a manifold. Since the convex open ball B'^o and F are contractible, N_1 and $p(F)$ are homotopy equivalent. If we double N_1 along $p(F)$, the the double is a closed orientable n-manifold homotopy equivalent to an $(n-1)$-manifold $p(F)$. This is absurd, and therefore, there are infinitely many copies of F in B.

By the conclusion of the above two paragraphs, given j, there exists i, $i > j$ with a deck transformation ϑ_i such that $\vartheta_i(F)$ is inside the component C_j of $B^o - \vartheta_j(F)$ not containing F. Let d_N denote the original Riemannian metric on $\tilde{N}$ induced from one of N. Then for each point x of F, $d_N(\vartheta_i(x), F) \to \infty$ since such a path from x to F has to pass through an increasing number of $\vartheta_j(F)$ as $i \to \infty$, and there exists a lower bound to the infimums of $d_N(y, z)$ if y and z are in two different components of $p^{-1}(F)$. (The latter statement follows since any arc in N with endpoints in $p(F)$ that is not homotopic to a path in $p(F)$ must have d_N-length longer than some uniform positive constant ϵ.)

Let K be any compact subset of $\tilde{N}$ intersecting F. Then it follows that $\vartheta_i(K)$ is eventually in B^o. From this we easily see that any two points y and z can be connected by $\vartheta_i^{-1}(s)$ for some i and a segment s in B^o of $\mathbf{d}$-length $\leq \pi$ since B^o is convex. Thus, $\tilde{N}$ is convex, and so is N. □

DEFINITION 12.3. Let R_1 and R_2 be both pseudo-crescent-cones or both infinite pseudo-crescent-cones. We say that R_1 and R_2 intersect transversally if the following statements hold ($i = 1, j = 2$, or $i = 2, j = 1$):

1. R_1 and R_2 overlap.
2. $\nu_{R_1} \cap \nu_{R_2}$ is a well-positioned radial segment s in ν_{R_1} and similarly in ν_{R_2}.
3. $\nu_{R_i} \cap R_j$ equals the closure of a component of $\nu_{R_i} - s$. Hence it is a well-positioned convex triangle D in R_j. Two sides of D other than s are finitely ideal and infinitely ideal respectively. The three sides of D are well-positioned in three sides of R_j respectively; or a finitely ideal side of D equals the common side of two finitely ideal sides of R_j and other two sides are well positioned in ν_{R_j} and an infinitely ideal sides of R_j respectively.
4. $R_i \cap R_j$ is the closure of a component of $R_i - D$.
5. Let t be the infinitely ideal side of D. $\alpha^{\mathsf{i}}_{R_i} \cap \alpha^{\mathsf{i}}_{R_j}$ is equal to one of two components of $\alpha^{\mathsf{i}}_{R_i} - t$. Hence, $\alpha^{\mathsf{i}}_{R_i} \cup \alpha^{\mathsf{i}}_{R_j}$ is an open 2-disk, i.e., $\alpha^{\mathsf{i}}_{R_i}$ and $\alpha^{\mathsf{i}}_{R_j}$ extend each other.
6. Let t' be the finitely ideal side of D. $\alpha^{\mathsf{f}}_{R_i} \cap \alpha^{\mathsf{f}}_{R_j}$ is equal to one of two components of $\alpha^{\mathsf{f}}_{R_i} - t'$. Hence, $\alpha^{\mathsf{f}}_{R_i} \cup \alpha^{\mathsf{f}}_{R_j}$ is an open 2-disk.

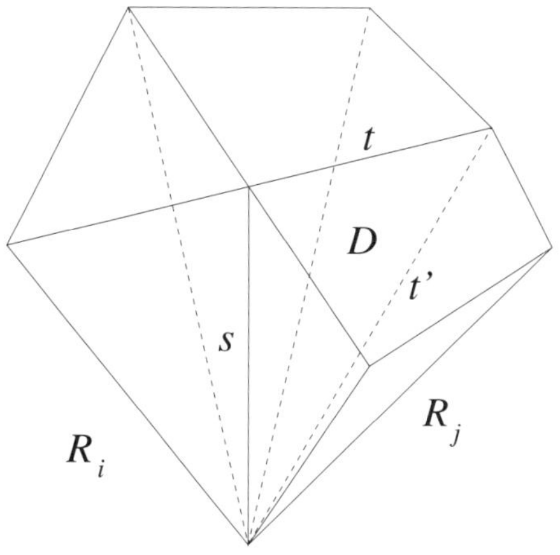

FIGURE 12.2. The transversal intersection of pseudo-crescent-cones.

7. $t \cap \alpha^{\mathsf{fi}}_{R_i}$ is a point. $\alpha^{\mathsf{fi}}_{R_i} \cap \alpha^{\mathsf{fi}}_{R_j}$ equals one of two components of $\alpha^{\mathsf{fi}}_{R_i} - t$, and $\alpha^{\mathsf{fi}}_{R_i} \cup \alpha^{\mathsf{fi}}_{R_j}$ is an open arc.

PROPOSITION 12.2. *Let R_1 and R_2 are both pseudo-crescent-cones overlapping with each other. Then either $R_1 \subset R_2$ or $R_2 \subset R_1$ holds or R_1 and R_2 intersect transversally.*

PROOF. The proof is similar to that of Theorem A.1. Only technical differences exist but we have to use the following lemma 12.3 instead of Lemma 5.2 in [**15**]. □

LEMMA 12.3. *Let N be a closed real projective manifold with a developing map* $\mathbf{dev} : \check{N}_h \rightarrow \mathbf{S}^3$ *for the completion $\check{N}_h$ of the holonomy cover N_h of N. Suppose that* **dev** *is an imbedding onto a union of two (infinite or otherwise) radiant convex polyhedra H_1 and H_2 meeting in an (infinite or otherwise) radiant polyhedron. Then $H_1 \cup H_2$ is convex and so is N.*

PROOF. Suppose that H_1 and H_2 are finite. If $H_1 \cup H_2$ is not convex, then there exist sides of H_1 and H_2 meeting in an edge so that they can be extend into H_1^o and H_2^o as totally geodesic surfaces respectively. We find a holonomy-group-invariant codimension-one submanifold in $H_1^o \cup H_2^o$ from these. The rest of the proof is similar to that of Lemma 5.2 in [**15**]. When at least one of H_1 and H_2 is infinite, a similar argument will work using support planes. □

PROPOSITION 12.3. *Let R_1 and R_2 be both infinite pseudo-crescent-cones overlapping with each other. Then either $R_1 \subset R_2$ or $R_2 \subset R_1$. The special edges of R_1 and R_2 must coincide in this case.*

PROOF. If the conclusion does not hold, then R_1 and R_2 must intersect transversally as the proof for Proposition 12.2 also works here. Since M is orientable, this means that a special edge e of the fundamental side of R_1 lies in $\mathcal{L}_2$ the union of finitely ideal sides of R_2 with the special edge of R_2 removed or conversely for R_1 and R_2. Assume the former case. The finitely ideal sides of R_1 near the special edge e must lie in $\mathcal{L}_2$ from the definition of transversality. However, the sides of R_2 are locally finite there while those of R_1 are not. This is a contradiction. The last statement follows obviously from the proof. □

PROPOSITION 12.4. *Let R_i be an ascending sequence of pseudo-crescent-cones or infinite pseudo-crescent-cones; that is, $R_1 \subset R_2 \subset R_3 \subset \dots$. Then there exists a pseudo-crescent-cone or an infinite pseudo-crescent-cone that includes the union of R_i as a dense subset.*

PROOF. By Theorem B.1, there exists a 3-ball R including every R_i. The fact that $\mathrm{Cl}(\alpha^{\mathsf{f}}_{R_i} \cup \alpha^{\mathsf{i}}_{R_i} \cup \alpha^{\mathsf{fi}}_{R_i})$ form an increasing ideal sequence shows that a part of the boundary of R is a subset of the ideal set. If $\check{M}$ is a subset of R, then $\tilde{M}$ equals R^o and M must be convex. Therefore, a side of the boundary of R is a subset of M, and R is the desired object. □

We will now show that infinite pseudo-crescent-cones do not exist on $\check{M}$. Suppose that R is an infinite pseudo-crescent-cone. We can introduce an equivalence relation on the collection $\mathcal{R}_2$ of all infinite pseudo-crescent-cones: We say two infinite pseudo-crescent-cones are equivalent if they overlap. The equivalence relation is generated by this; i.e., $R \sim S$ if there exists a finite sequence $R^1, R^2, \dots, R^n$ in $\mathcal{R}_2$ such that $R^1 = R$ and $R^n = S$ where R^i and R^{i+1} overlap for each $i = 1, \dots, n-1$.

Propositions 12.3 and 12.4 show that the equivalence class of overlapping infinite pseudo-crescent-cones is totally ordered and has a unique maximal element. Let R be the maximal element. For a deck transformation ϑ, either R and $\vartheta(R)$ do not overlap or $R \subset \vartheta(R)$ or $\vartheta(R) \subset R$ by Proposition 12.3. If $R \subset \vartheta(R)$, then since R is maximal, we have $R = \vartheta(R)$; if $\vartheta(R) \subset R$, then we have $R = \vartheta(R)$. If R and $\vartheta(R)$ do not overlap but meet each other, then $R \cap \vartheta(R)$ is the union of common components of $\nu_R \cap \tilde{M}$ and $\nu_{\vartheta(R)} \cap \tilde{M}$, which follows as in Chapter 7 of [**15**]. As in Chapter 10 of [**15**], we obtain the so-called two-faced submanifolds. If a two-faced submanifold exists, then a component of $\nu_R \cap \tilde{M}$ covers a closed surface in M, i.e., the two-faced submanifold. This was ruled out by Proposition 12.1, and we have either $R = \vartheta(R)$ or $R \cap \vartheta(R) \cap \tilde{M} = \emptyset$. (The detail of a similar argument was in Chapters 4 and 10 and we will see it again in this chapter.)

By Proposition 10.3, the collection consisting of elements of form $\vartheta(R)$ is locally finite in $\tilde{M}$. The equivariance and this show by Proposition 2.1 that R covers a compact submanifold in M. Again this means that a component of $\nu_R \cap \tilde{M}$ covers a closed surface in M, a contradiction. Therefore, we proved that there exist no infinite pseudo-crescent-cones:

We restate our result as follows:

PROPOSITION 12.5. *Let N be a compact radiant affine manifold with empty or totally geodesic boundary. If $\check{N}$ includes infinite pseudo-crescent cones, then N is convex.*

We will now show that pseudo-crescent-cones do not exist as well using a longer more involved argument. Let $\mathcal{R}_3$ be a collection of all pseudo-crescent-cones in $\check{M}$. We define a relation that $R \sim S$ for $R, S \in \mathcal{R}_3$ if they overlap, and define an equivalence relation on $\mathcal{R}_3$ generated by this as usual.

We define the following sets as in [**11**]:

$$\Lambda^{\mathsf{p}}(R) = \bigcup_{S\sim R} S, \quad \delta^{\mathsf{i}}_\infty\Lambda^{\mathsf{p}}(R) = \bigcup_{S\sim R} \alpha^{\mathsf{i}}_S, \quad \delta^{\mathsf{f}}_\infty\Lambda^{\mathsf{p}}(R) = \bigcup_{S\sim R} \alpha^{\mathsf{f}}_S,$$
$$\delta^{\mathsf{fi}}_\infty\Lambda^{\mathsf{p}}(R) = \bigcup_{S\sim R} \alpha^{\mathsf{fi}}_S, \quad \Lambda^{\mathsf{p}}_1(R) = \bigcup_{S\sim R} S - \nu_S \quad \text{for } R \in \mathcal{R}_3. \tag{12.1}$$

We begin by introducing the equivariance properties of $\Lambda^{\mathsf{p}}(R)$. We list properties of $\Lambda^{\mathsf{p}}(R)$.

1. $\Lambda^{\mathsf{p}}(R)$ and $\Lambda^{\mathsf{p}}_1(R)$ are path-connected.

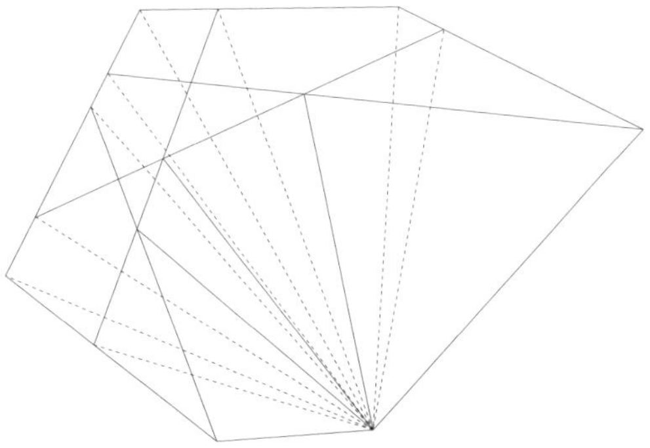

FIGURE 12.3. A picture of $\Lambda^{\mathsf{p}}(R)$.

2. $\Lambda^{\mathsf{p}}(R) \cap \tilde{M}$ is a closed radiant subset of $\tilde{M}$ with totally geodesic boundary, which is also a radiant set.
3. $\Lambda^{\mathsf{p}}_1(R)$ is a 3-manifold with boundary (and corners) $\delta_\infty \Lambda^{\mathsf{p}}(R)$, defined as $\delta^{\mathsf{i}}_\infty \Lambda^{\mathsf{p}}(R) \cup \delta^{\mathsf{f}}_\infty \Lambda^{\mathsf{p}}(R) \cup \delta^{\mathsf{fi}}_\infty \Lambda^{\mathsf{p}}(R)$. The 3-manifold has a real projective structure induced from the pseudo-crescent-cones.
4. $\delta^{\mathsf{i}}_\infty \Lambda^{\mathsf{p}}(R)$ is an open surface imbedded in $\tilde{M}^{\mathsf{i}}_\infty$.
5. $\delta^{\mathsf{f}}_\infty \Lambda^{\mathsf{p}}(R)$ is an open surface imbedded in $\tilde{M}_\infty - \tilde{M}^{\mathsf{i}}_\infty$.
6. $\delta^{\mathsf{fi}}_\infty \Lambda^{\mathsf{p}}(R)$ is an arc in $\tilde{M}^{\mathsf{i}}_\infty$, and the union $\delta_\infty \Lambda^{\mathsf{p}}(R)$ is an open surface imbedded in $\tilde{M}_\infty$.
7. $\Lambda^{\mathsf{p}}(R)$ is *maximal*, i.e., for any triangle T in $\tilde{M}$ with sides s_1, s_2, and s_3, if s_2 and s_3 are subsets of $\Lambda^{\mathsf{p}}(R)$, then s_1 is a subset of $\Lambda^{\mathsf{p}}(R)$.
8. $\mathrm{bd}\Lambda^{\mathsf{p}}(R) \cap \tilde{M}$ is a properly imbedded countable union of disks in $\tilde{M}^o$, each of which is a totally geodesic, radiant, properly imbedded open triangles.
9. For any deck transformation ϑ, we have

$$\Lambda^{\mathsf{p}}(\vartheta(R)) = \vartheta(\Lambda^{\mathsf{p}}(R)), \qquad \delta^{\mathsf{i}}_\infty \Lambda^{\mathsf{p}}(\vartheta(R)) = \vartheta(\delta^{\mathsf{i}}_\infty \Lambda^{\mathsf{p}}(R)),$$
$$\delta^{\mathsf{f}}_\infty \Lambda^{\mathsf{p}}(\vartheta(R)) = \vartheta(\delta^{\mathsf{f}}_\infty \Lambda^{\mathsf{p}}(R)), \quad \delta^{\mathsf{fi}}_\infty \Lambda^{\mathsf{p}}(\vartheta(R)) = \vartheta(\delta^{\mathsf{fi}}_\infty \Lambda^{\mathsf{p}}(R)).$$

The proof of these facts are similar to what is given in [**11**] for dimension two and Chapter 7 of [**15**]. Note that for the closedness in (2), we use the fact that a sequence of pseudo-crescent-cones with a common open set in them give rise to a pseudo-crescent-cone or infinite pseudo-crescent-cone by Theorem B.1. Since infinite pseudo-crescent cones do not exist, we obtain the closedness in (2).

PROPOSITION 12.6. *Given two pseudo-crescent-cones R and S, we have one of three possibilities*:

- $\Lambda^{\mathsf{p}}(R) = \Lambda^{\mathsf{p}}(S)$,
- $\Lambda^{\mathsf{p}}(R)$ *meets* $\Lambda^{\mathsf{p}}(S)$ *at the union of common components of* $\mathrm{bd}\Lambda^{\mathsf{p}}(R) \cap \tilde{M}$ *and* $\mathrm{bd}\Lambda^{\mathsf{p}}(S) \cap \tilde{M}$.
- $\Lambda^{\mathsf{p}}(R) \cap \tilde{M}$ *and* $\Lambda^{\mathsf{p}}(S) \cap \tilde{M}$ *are disjoint.*

PROOF. The proof is same as in Chapter 6 or 7 of [**15**]. The role played by n-crescent is played by pseudo-crescent-cones. □

PROPOSITION 12.7. $\delta^{\mathsf{f}}_\infty \Lambda^{\mathsf{p}}(R) \cup \delta^{\mathsf{fi}}_\infty \Lambda^{\mathsf{p}}(R) \cup \{O\}$ *is a union of countable collection of triangles T_i. Each T_i is a side of a pseudo-crescent-cone S, $S \sim R$ in the closure of α^{f}_S. Precisely one of the following holds*:

1. *The collection equals $\{T_0, T_1, \dots, T_{n-1}\}$ for $n \geq 2$ and T_i and T_j meet in an edge if $i - j = 1, n-1 \bmod n$ and T_i and T_j are disjoint if otherwise and $i \neq j$. (Indices are cyclic.)*
2. *The collection equals $\{T_0, T_1, \dots, T_{n-1}, \dots\}$ for $n \geq 2$ and T_i and T_j meet in an edge if $i - j = \pm 1$ and T_i and T_j are disjoint if otherwise and $i \neq j$. (The collection could be finite or infinite).*
3. *The collection equals $\{T_i | i \in \mathbf{Z}\}$ and T_i and T_j meet in an edge if $i - j = \pm 1$ and T_i and T_j are disjoint if otherwise and $i \neq j$.*

PROOF. This follows from Proposition 12.2. □

Let R be a pseudo-crescent-cone. We orient each radial segment toward the origin O. The collection of finitely ideal sides of R in $\tilde{M}^{\mathsf{f}}_\infty$ can be linearly ordered using the boundary orientation induced from that of R, which in turn was obtained from $\tilde{M}$: We give the orientation on $\tilde{M}$ and the ordering on $\{T_i\}$ so that T_{i+1} is to the right of T_i for each $i = 0, 1, \ldots$.

The *right-most* ideal side of a pseudo-crescent-cone R is the finitely ideal side of R which is right-most among the finitely ideal sides of R.

LEMMA 12.4. *Each T_i is the right-most ideal side of the unique pseudo-crescent-cone maximal among pseudo-crescents having T_i as the right-most ideal side.* (*In the second case above we have to assume $i \geq 2$.*)

PROOF. Clearly, there exists at least one pseudo-crescent-cone with the right-most side T_i since we can choose such a pseudo-crescent-cone in a pseudo-crescent-cone S, $S \sim R$, including T_i as a side in the closure of α^{f}_S. Given any two such pseudo-crescent cones R and S, since they overlap, we have either $R \subset S$ or $S \subset R$ or R and S intersect transversally by Proposition 12.2. If R and S intersect transversally, we can see easily that R and S cannot share the right-most ideal side. Hence, we have $R \subset S$ or $S \subset R$.

Also, given any ascending sequence $R_{i,j}$ of pseudo-crescent-cones such that T_i is the right-most ideal side, we see that $\bigcup_j R_{i,j}$ is contained in a pseudo-crescent-cone by Proposition 12.4, which has T_i as the right-most ideal side. The uniqueness of the maximal element follows easily from the previous paragraph. □

We rule out the first possibility for the triangles.

PROPOSITION 12.8. *The collection $\{T_i\}$ is linearly ordered.*

PROOF. Suppose not. Then we obtain by Lemma 12.4 a sequence of pseudo-crescent-cones P_i such that P_i and P_{i+1} overlap for $i = 0, \ldots, n-1 \bmod n$ where the indices are cyclic. Orient ν_{P_i} by the outer-normal to P_i for each i. $\nu_{P_i} \cap P_{i+1}$ is the closure of the left component of $\nu_{P_i} - \nu_{P_{i+1}}$ and $\nu_{P_i} \cap P_{i-1}$ the closure of the right component of $\nu_{P_i} - \nu_{P_{i-1}}$. We see that $\nu_{P_i} \cap P^o_{i-1}$ is contained the right-most component of $\nu^o_{P_i} - \tilde{M}^{\mathrm{f}}_\infty$ and $\nu_{P_i} \cap P^o_{i+1}$ is contained in the left-most component of $\nu^o_{P_i} - \tilde{M}^{\mathrm{f}}_\infty$. Hence, we may choose a closed simple arc α in $\tilde{M}$ so that $\alpha \cap P_i$ is a connected arc with two endpoints respectively in $\nu_{P_i} \cap P^o_{i-1}$ and $\nu_{P_i} \cap P^o_{i+1}$ for each i.

Suppose that $\nu^o_{P_i} \cap \tilde{M}^{\mathrm{f}}_\infty$ is empty. Assume first that $\nu^o_{P_i} - P^o_{i-1} - P^o_{i+1}$ is not empty. Then there exists a segment s in $\nu^o_{P_i} - P^o_{i-1} - P^o_{i+1}$. The segment has a compact neighborhood N in $\tilde{M}$ which includes another segment s' in $\tilde{M}$ outside P_i ending at $p' \in \nu_{P_{i-1}}$ and $q' \in \nu_{P_{i+1}}$ transversally. We extend s' in P_{i-1} and P_{i+1} to a segment s'' with endpoints in $p'' \in \alpha^{\mathrm{f}}_{P_{i-1}}$ and $q'' \in \alpha^{\mathrm{f}}_{P_{i+1}}$ respectively where we assume that points occur on s'' in order p'', p', q', q''. We are additionally required to choose s'' so that $q'' \in \delta\nu_{P_i} \cap T_i$. We may also assume that s and s' are the sides of a totally geodesic convex disk D with four edges where two other edges are in $\nu_{P_{i-1}}$ and $\nu_{P_{i+1}}$ respectively. Then the union of all radial segments passing through D is a convex 3-ball K bounded by five sides. The union of K, P_{i-1}, P_i, and P_{i+1} contains a pseudo-crescent cone containing $P_i \cup K$ properly and having T_i as its right-most ideal side. To see this, we need to make K slightly larger and apply Proposition 1.1 as in the proof of Lemma 5.1. This is a contradiction since P_i is maximal.

If $\nu^o_{P_i} - P^o_{i-1} - P^o_{i+1}$ is empty, then $\nu^o_{P_i}$ is a subset of $P^o_{i-1} \cup P^o_{i+1}$. We clearly see that there exists a radiant triangle in $P_{i-1} \cup P_{i+1}$ near ν_{P_i} outside P_i which bounds a pseudo-crescent cone P'_i including P_i. We can choose P'_i so that T_i is the right-most ideal side of P_i'. This is a contradiction. Therefore, $\nu^o_{P_i} \cap \tilde{M}^{\mathsf{f}}_\infty$ is not empty.

We may now choose a properly imbedded radiant open triangle $\triangle_i$ in P_i^o such that $\triangle_i$ meets α at a unique point (up to isotopying α in P^o), and $\mathrm{Cl}(\triangle_i) - \triangle_i$ is the union of a radial segment in $\nu_{P_i} \cap M_{h\infty}$, a radial segment in $\mathrm{Cl}(\alpha^{\mathsf{f}}_{P_i})$, and a segment in $\mathrm{Cl}(\alpha^{\mathsf{i}}_{P_i})$ and is a subset of $\tilde{M}_\infty$. Since $\tilde{M}$ is simply connected, α must bound a disk D. Perturbing D and α into general positions with respect to $\triangle_i$ makes $D \cap \triangle_i$ to a properly imbedded arcs in D with endpoints in δD since $\triangle_i$ is properly imbedded; thus, α intersects with $\triangle_i$ even number of times by the classification of 1-manifolds, a contradiction. □

CLAIM 12.1. *$\tilde{M}$ is a subset of $\Lambda^{\mathsf{p}}(R)$, and M is boundaryless.*

PROOF. Suppose that $\Lambda^{\mathsf{p}}(R) \cap \tilde{M}$ is a proper subset of $\tilde{M}$; i.e., $\mathrm{bd}\Lambda^{\mathsf{p}}(R) \cap \tilde{M}$ is not empty. For any pair of pseudo-crescent-cones R and S, one of the following possibilities hold from Proposition 12.6:

- $\Lambda^{\mathsf{p}}(R) = \Lambda^{\mathsf{p}}(S)$,
- $\Lambda^{\mathsf{p}}(R) \cap \Lambda^{\mathsf{p}}(S) \cap \tilde{M} = \emptyset$ where $R \not\sim S$, or
- $\Lambda^{\mathsf{p}}(R) \cap \tilde{M}$ and $\Lambda^{\mathsf{p}}(S) \cap \tilde{M}$ meet at the union of common components of $\mathrm{bd}\Lambda^{\mathsf{p}}(R) \cap \tilde{M}$ and $\mathrm{bd}\Lambda^{\mathsf{p}}(S) \cap \tilde{M}$ where $R \not\sim S$. As in [**15**], this set is also the union of common components of $\nu_T \cap \tilde{M}$ for a pseudo-crescent-cone T, $T \sim R$ and $\nu_U \cap \tilde{M}$ for $U \sim S$.

The third possibility gives rise to a closed submanifold of codimension 1 as in [**15**] or in above Chapters 4 and 10, analogously to the two-faced manifolds. This means that a component of $\nu_T \cap \tilde{M}$ covers a compact submanifold for a pseudo-crescent-cone T, which is a contradiction by Proposition 12.1, and the third possibility does not occur.

Since for every deck transformation ϑ, $\vartheta(\Lambda^{\mathsf{p}}(R)) = \Lambda^{\mathsf{p}}(\vartheta(R))$, and there is no two-faced submanifold, we have that $\Lambda^{\mathsf{p}}(R) = \vartheta(\Lambda^{\mathsf{p}}(R))$ or $\Lambda^{\mathsf{p}}(R) \cap \vartheta(\Lambda^{\mathsf{p}}(R)) \cap \tilde{M} = \emptyset$.

The collection consisting of elements of form $\vartheta(\Lambda^{\mathsf{p}}(R))$ is locally finite since this can be proved by the proof of Proposition 10.2 replacing crescent-cones by pseudo-crescent-cones. By Proposition 2.1, we have that $\Lambda^{\mathsf{p}}(R) \cap \tilde{M}$ covers a compact submanifold N in M. Recall that $\mathrm{bd}\Lambda^{\mathsf{p}}(R) \cap \tilde{M}$ is totally geodesic. Let F be a component of $\mathrm{bd}\Lambda^{\mathsf{p}}(R)$. Then a point x of F belongs to ν_T where T is a radiant bihedron equivalent to R. Since $\nu_T \cap \tilde{M}$ and F must be tangent there, and F is properly imbedded, it follows that F is a component of $\nu_T \cap \tilde{M}$. Since F covers a component of the boundary of N, F covers a closed surface in N, which was ruled out by Proposition 12.1, and our claim that $\tilde{M}$ is a subset of $\Lambda^{\mathsf{p}}(R)$ is proved.

Suppose that δM is not empty. There exists a component K of $\delta\tilde{M}$. Let x be a point of K. Then x belongs to $\nu_T \cap \tilde{M}$ for some $T \sim R$. Let K' be the component of $\nu_T \cap \tilde{M}$ containing x. Then since K and K' must be tangent at x, and both are properly imbedded, we have $K = K'$. However, Proposition 12.1 shows that this is a contradiction. □

As before, $\vartheta(\Lambda^{\mathsf{p}}(R)) = \Lambda^{\mathsf{p}}(\vartheta(R))$ for a deck transformation ϑ includes $\tilde{M}$ and hence $\vartheta(\Lambda^{\mathsf{p}}(R))$ meets $\Lambda^{\mathsf{p}}(R)^o$. Hence

$$\vartheta(\Lambda^{\mathsf{p}}(R)) = \Lambda^{\mathsf{p}}(R), \quad \vartheta(\delta^{\mathsf{f}}_\infty \Lambda^{\mathsf{p}}(R)) = \delta^{\mathsf{f}}_\infty \Lambda^{\mathsf{p}}(R),$$
$$\vartheta(\delta^{\mathsf{i}}_\infty \Lambda^{\mathsf{p}}(R)) = \delta^{\mathsf{i}}_\infty \Lambda^{\mathsf{p}}(R), \quad \vartheta(\delta^{\mathsf{fi}}_\infty \Lambda^{\mathsf{p}}(R)) = \delta^{\mathsf{fi}}_\infty \Lambda^{\mathsf{p}}(R)$$

hold for every deck transformation ϑ. We have that the deck transformation group $\pi_1(M)$ of $\check{M}$ acts on the discrete ordered set $\{T_i\}$ either preserving the order or reversing the order since adjacent triangles have to go adjacent ones. We have an exact sequence:

$$1 \to K \to \pi_1(M) \to Q \to 1$$

where Q is the group of automorphisms of linearly ordered set $\{T_i\}$ and K is the kernel. Elements of Q may preserve the order or reverse it. We assume without loss of generality that Q preserves the order (by choosing a double cover of M if necessary).

Suppose that Q is a trivial group; that is, $\pi_1(M)$ acts trivially on $\{T_i\}$. Then $\pi_1(M)$ acts on each of T_i. For $i \geq 2$, choose a maximal pseudo-crescent-cone R which contains T_i as the right-most ideal side. Then for each deck transformation ϑ, $\vartheta(R)$ is the maximal such pseudo-crescent-cone, and hence $R = \vartheta(R)$. Since $\pi_1(M)$ acts on R, the submanifold $R \cap \check{M}$ covers a compact submanifold of M with boundary union of closed surfaces by Proposition 2.1. This shows that a component $\nu_R \cap \check{M}$ covers a closed surface in M, a contradiction to Proposition 12.1.

Suppose now that the action is not trivial so that Q is isomorphic to $\mathbf{Z}$. By Theorem 11.6 in Hempel [**26**], we see that the Poincaré associate of M is a fiber bundle over a circle, and K is nothing but the fundamental group of the fiber which must be a closed surface as M is a closed 3-manifold. Hence, the subgroup K of $\pi_1(M)$ isomorphic to the fundamental group of a surface acts trivially on $\{T_i\}$. As above choose a maximal pseudo-crescent-cone R_i which contains T_i as the right-most ideal side. Then K acts on $\nu_R \cap \check{M}$, a union of disjoint open sets.

Since the action of $\pi_1(M)$ on the collection T_i is not trivial, the set of indices of T_i equals $\mathbf{Z}$; so we choose a fixed i. R_{i+1} exists, and since R_i and R_{i+1} are maximal and R_i does not include T_{i+1}, either $R_i \subset R_{i+1}$ or R_i and R_{i+1} intersect transversally. In the first case, since R_i is a proper subset of R_{i+1}, it follows from looking at their images under **dev** that $\nu^o_{R_i} \subset R^o_{i+1}$. In the second case, R_i and R_{i+1} intersect transversally, and $\nu_{R_i} \cap R^o_{i+1}$ is a radiant open triangle adjacent to T_i. In both cases, let E denote the open triangle $\nu_{R_i} \cap R^o_{i+1}$.

Since K acts on T_i, K acts on the component L of $\nu_{R_i} \cap \check{M}$ including E. Since K is isomorphic to the fundamental group of a closed surface F and L is homeomorphic to an open disk, it follows that L/K is a closed surface homeomorphic to F, contradicting Proposition 12.1 as above. Therefore, we completed to show that there exists no pseudo-crescent in $\check{M}$ and hence in $\check{M}_h$ by Lemma 12.1.

APPENDIX A

Dipping intersections

We will gather needed facts about the dipping intersection of n-balls and the transversal intersection of n-crescents here, which were already stated and proved in [**11**] and [**15**].

Let D be a convex n-ball in $\check{M}_h$ such that δD includes a totally geodesic convex $(n-1)$-ball α. We say that a convex n-ball F is *dipped into* (D, α) if the following statements hold:

- D and F overlap.
- $F \cap \alpha$ is a tame $(n-1)$-ball β with $\delta\beta \subset \delta F$ and $\beta^o \subset F^o$.
- $F - \beta$ has two convex components O_1 and O_2 such that $\mathrm{Cl}(O_1) = O_1 \cup \beta = F - O_2$ and $\mathrm{Cl}(O_2) = O_2 \cup \beta = F - O_1$.
- $F \cap D$ is equal to $\mathrm{Cl}(O_1)$ or $\mathrm{Cl}(O_2)$.

We say that F is *dipped into* (D, α) *nicely* if the following statements hold:

- F is dipped into (D, α).
- $F \cap D^o$ is identical with O_1 and O_2.
- $\delta(F \cap D) = \beta \cup \xi$ for a topological $(n-1)$-ball ξ in the topological boundary $\mathrm{bd}F$ of F in $\check{M}$ where $\beta \cap \xi = \delta\beta$.

As a consequence, we have $\delta\beta \subset \mathrm{bd}F$.

The direct generalization of Corollary 1.9 of [**11**] gives:

COROLLARY A.1. *Suppose that F and D overlap, and $F^o \cap (\delta D - \alpha^o) = \emptyset$. Assume the following two equivalent conditions:*

- $F^o \cap \alpha \neq \emptyset$.
- $F \not\subset D$.

Then F is dipped into (D, α). If $F \cap (\delta D - \alpha^o) = \emptyset$ furthermore, then F is dipped into (D, α) nicely.

Suppose that R_1 is an n-crescent that is an n-bihedron. Let R_2 be another bihedral n-crescents with sets α_{R_2} and ν_{R_2}. We say that R_1 and R_2 intersect *transversally* if the following conditions hold ($i = 1, j = 2$; or $i = 2, j = 1$):

1. $\nu_{R_1} \cap \nu_{R_2}$ is an $(n-2)$-dimensional hemisphere if $n > 2$ and consists of a single point x if $n = 2$.
2. For the intersection $\nu_{R_1} \cap \nu_{R_2}$ denoted by H, H^o is a subset of the interior $\nu^o_{R_i}$, and $\nu^o_{R_i}$ and $\nu^o_{R_j}$ intersect transversally at points of H^o.
3. $\nu_{R_i} \cap R_j$ is a tame $(n-1)$-bihedron with boundary the union of H and an $(n-2)$-hemisphere H' in the closure of α_{R_j} with its interior H'^o in α_{R_j}.
4. $\nu_{R_i} \cap R_j$ is the closure of a component of $\nu_{R_i} - H$ in $\check{M}$.
5. $R_i \cap R_j$ is the closure of a component of $R_j - \nu_{R_i}$.
6. Both $\alpha_{R_i} \cap \alpha_{R_j}$ and $\alpha_{R_i} \cup \alpha_{R_j}$ are homeomorphic to open $(n-1)$-dimensional balls. (See Figure A.1.)

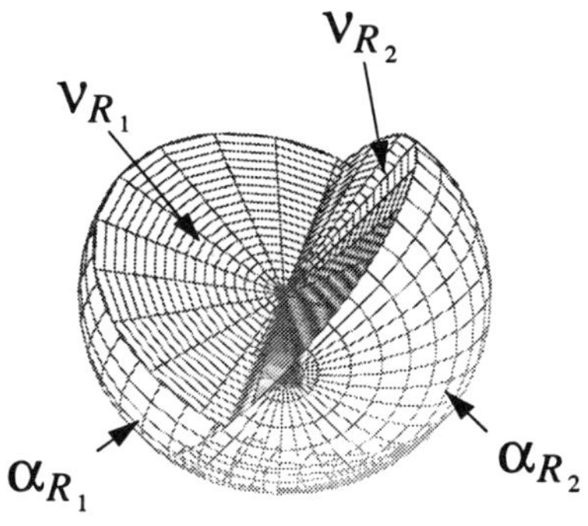

FIGURE A.1. A transversal intersection of 3-crescents (stereographically projected to $\mathbf{R}^3$).

The above mirrors the property of intersection of $\mathbf{dev}(R_1)$ and $\mathbf{dev}(R_2)$ where $\mathbf{dev}(\alpha_{R_1})$ and $\mathbf{dev}(\alpha_{R_2})$ are included in a common great sphere $\mathbf{S}^{n-1}$ of dimension $(n-1)$, $\mathbf{dev}(R_1)$ and $\mathbf{dev}(R_2)$ are included in a common n-hemisphere bounded by $\mathbf{S}^{n-1}$ and $\mathbf{dev}(\nu_{R_1})^o$ and $\mathbf{dev}(\nu_{R_2})^o$ meet transversally (see Proposition 1.1).

THEOREM A.1 (Theorem 5.4 [**15**]). *Suppose that R_1 and R_2 are overlapping. Then either R_1 and R_2 intersect transversally, or $R_1 \subset R_2$ or $R_2 \subset R_1$ hold.*

PROOF. The proof is in [**15**]. Note we need the following lemma to prove this. □

LEMMA A.1 (Lemma 5.2 [**15**]). *Let N be a closed real projective n-manifold. Suppose that $\mathbf{dev} : \check{N}_h \to \mathbf{S}^n$ is an imbedding onto the union of n-hemispheres H_1 and H_2 meeting each other in an n-bihedron or an n-hemisphere. Then $H_1 = H_2$, and N_h is projectively diffeomorphic to an open n-hemisphere.*

APPENDIX B

Sequences of n-balls

First, we discuss for $\mathbf{S}^n$ the convergence of a sequence of convex n-balls, the limit of a convergent sequence, and the boundary and the interior of the limit. Next, we consider sequences of convex n-balls in a Kuiper completion. Subsequences may not converge since the Kuiper completion is not compact in general. However, when a sequence has a *core*, i.e., a common convex open n-ball in every sequence element, a subsequence will have a "limit." So, a certain criterion assuring the existence of a core is presented first. Finally, certain sequences with cores are shown to "converge." The details of this chapter is given in Part I of [**15**].

Since $\mathbf{S}^n$ is a compact metric space, the collection of all compact subsets of $\mathbf{S}^n$ form a compact metric space with the *Hausdorff* metric $\mathbf{d}^H$ defined as follows: for compact subsets A and B, given $\epsilon > 0$, we define $\mathbf{d}^H(A, B) < \epsilon$ if A is in ϵ-$\mathbf{d}$-neighborhood of B and B is in that of A.

A sequence of compact convex subsets of $\mathbf{S}^n$ always has a convergent subsequence $\{D_i\}$ with respect to $\mathbf{d}$. The limit is always a compact convex subset. The dimension of the limit is less than or equal to $\liminf_{i\rightarrow\infty} \dim(D_i)$. We deduce from this that the limit of a convergent sequence of convex n-balls is a convex m-ball for $m \leq n$. As in [**11**], if $m = n$, we can deduce that δD is the limit of $\{\delta D_i\}$, and $\bigcup_{i=1}^{\infty} D_i^o \supset D^o$ (see Part I of [**15**]).

LEMMA B.1. *Let D_i be a sequence of 3-balls in $\mathrm{Cl}(\mathcal{H})$ which are cones over 2-disks B_i in $\mathbf{S}^2_\infty$. Then D_i converges to a limit compact set D_∞ if and only if B_i converges to a limit compact set B_∞ in $\mathbf{S}^2_\infty$. Moreover, in this case, D_∞ is a cone over B_∞.*

PROOF. Straightforward. □

Recall that μ denote the Riemannian metric on $\check{M}_h$ induced from that of $\mathbf{S}^n$ by **dev**.

LEMMA B.2 (Proposition 3.14 [**15**]). *Let $\{D_i\}$ be a sequence of convex n-balls in $\check{M}_h$. Let $x \in M_h$, and $B(x)$ a tiny ball of x. Suppose that the following properties hold:*

1. *δD_i includes an $(n-1)$-ball ν_i.*
2. *$B(x)$ overlaps with D_i and does not meet $\delta D_i - \mathrm{Cl}(\nu_i)$.*
3. *A sequence $\{x_i\}$ converges to x where $x_i \in \nu_i$ for each i.*
4. *The sequence $\{\mathbf{n}_i\}$ converges where $\mathbf{n}_i$ is the outer-normal vector to ν_i at x_i with respect to μ for each i.*

Then there exist a positive integer N and a convex open disk $\mathcal{P}$ in $B(x)$ such that

$$\mathcal{P} \subset D_i \text{ whenever } i > N.$$

We say that a compact subset D_∞ of $\mathbf{S}^n$ is the *resulting set* of a sequence $\{D_i\}$ of compact subsets of $\check{M}_h$ if $\{\mathbf{dev}(D_i)\}$ converges to D_∞. Let $\{D_i\}$ and $\{B_i\}$ be sequences of convex n-balls with resulting sets D_∞ and B_∞ respectively; let $\{K_i\}$ be a sequence of compact subsets with the resulting set K_∞. We say that $\{D_i\}$ *subjugates* $\{K_i\}$ if $D_i \supset K_i$ for each i and that $\{B_i\}$ *dominates* $\{D_i\}$ if B_i and D_i overlap for each i and if $B_\infty \supset D_\infty$. Moreover, we say that $\{K_i\}$ is *ideal* if there is a positive integer N for every compact subset F of M_h such that $F \cap K_i = \emptyset$ whenever $i > N$. In particular, if K_i is a subset of $M_{h\infty}$ for each i, then $\{K_i\}$ is an ideal subjugated sequence.

THEOREM B.1 (Proposition 3.15 [**15**]). *Suppose that $\{D_i\}$ is a sequence of n-balls including a common open ball $\mathcal{P}$, $\{D_i\}$ subjugates a sequence of compact subsets $\{K_i\}$, and a sequence of n-balls $\{B_i\}$ dominates $\{D_i\}$. Let D_∞, B_∞, and K_∞ be the resulting sets of the above sequences $\{D_i\}, \{B_i\}$, and $\{K_i\}$. Then $\check{M}_h$ includes two convex n-balls D^u and B^u and a compact subset K^u with the following properties*:

1. *$D^u \supset \mathcal{P}$, and $\mathbf{dev}(D^u) = D_\infty$.*
2. *$B^u \supset D^u$, and $\mathbf{dev}(B^u) = B_\infty$.*
3. *$D^u \supset K^u$, and $\mathbf{dev}(K^u) = K_\infty$.*
4. *If $\{K_i\}$ is ideal, then $K^u \subset M_{h\infty}$.*

APPENDIX C

Radiant affine 3-manifolds with boundary, and certain radiant affine 3-manifolds

By Thierry Barbot and Suhyoung Choi

Let G be a group acting on an analytic manifold X. An (X, G)-manifold is a manifold admitting an atlas with charts with value in X and whose coordinate change mappings are restrictions of elements of G. It is well-known that equipping a manifold M with an (X, G)-structure is equivalent to giving a pair $(\mathbf{dev}, \rho)$, where $\mathbf{dev}$ is an immersion from the universal covering $\tilde{M}$ of M into X, and where ρ is a homomorphism from the fundamental group Γ of M into G, such that

$$\forall \gamma \in \Gamma \quad \mathbf{dev} \circ \gamma = \rho(\gamma) \circ \mathbf{dev}.$$

Here, the action of Γ on $\tilde{M}$ is of course the action by deck transformations. The map $\mathbf{dev} : \tilde{M} \to X$ is the *developing map* of the structure, and $\rho : \Gamma \to G$ is the *holonomy homomorphism.*

A radiant affine n-manifold is an (X, G)-manifold, where X is the vector space $\mathbf{R}^n$, and G is the group $\mathrm{GL}(n, \mathbf{R})$ of linear transformations (see [**10**]). Such a manifold is naturally equipped with a transversely projective flow, the so-called *radial flow,* defined as follows: if $(x_1, \ldots, x_n)$ are local coordinates, the vector field generating the radial flow is:

$$X(x_1, \ldots, x_n) = \Sigma_{i=1}^{n} x_i \partial_{x_i}$$

Observe that this vector field does not depend on the coordinate systems as long as the origins are the same, and thus induces well-defined vector field on $\tilde{M}$ and on M, which is said to be a *radiant vector field.* The flow generated by the vector field is said to be a *radial flow.* The radial flow in $\mathbf{R}^n$ has unique singularity at the origin O but the radial flow on M has no singularity since $\mathbf{dev}$ misses O (see [**10**] and Lemma 3.1).

Let N be a closed real projective $(n-1)$-manifold, i.e. an $(\mathbf{R}P^{n-1}, \mathrm{PGL}(n, \mathbf{R})))$-manifold, where $\mathbf{R}P^{n-1}$ is the real projective space of dimension $n-1$, and $\mathrm{PGL}(n, \mathbf{R})$ is the group of projective transformations. Let φ be a projective automorphism of N. We can associate to the pair (N, φ) a family of radiant affine closed n-manifold: i.e., generalized affine suspensions, homeomorphic to a topological suspension of N by φ (see Chapter 3 and [**7, 42, 10, 4**]). They can be characterized by the following property: *a closed radiant affine manifold is a generalized affine suspension if and only if its radial flow admits a total cross-section, i.e. there is a closed embedded submanifold transverse everywhere to the flow and which meets every orbit of the flow.* (See Proposition 3.2.) (Note that the term "affine suspension is reserved for the case when N and φ are both affine.) A *Benzécri suspension* is an affine suspension so that φ is the identity or a finite-order automorphism of N. In this case all orbits are closed, and N is Seifert-fibered.

In this appendix, we study the following particular case which was left from Chapter 11, i.e., Lemma 11.6, which is implied by the following theorem since developing maps of a universal cover is always obtainable from developing maps of the holonomy cover composed with the covering map to the holonomy cover from the universal cover.

THEOREM C.1. *There is no closed radiant affine 3-manifold whose developing map from the universal cover is an infinite cyclic covering over* $\mathbf{R}^3$ *minus a line.*

A subsurface S in an affine 3-manifold is *totally-geodesic* if every point of S has a neighborhood $\mathcal{O}$ affinely diffeomorphic to an open subset of $\mathbf{R}^3$ or of an affine half-space of $\mathbf{R}^3$ so that $S \cap \mathcal{O}$ corresponds to a closed affine subspace of codimension-one intersected with the open set. A totally geodesic subsurface has a natural induced affine structure as a two-dimensional manifold. A boundary component of an affine 3-manifold is *totally geodesic* if each boundary point has a neighborhood affinely diffeomorphic to an open subset of a half-space in $\mathbf{R}^3$.

THEOREM C.2 (Theorem B of Barbot [**3**]). *Let M be a closed radiant affine 3-manifold. If M includes a totally geodesic surface tangent to the radial flow, then M admits a total cross-section; i.e., M is a generalized affine suspension.*

The second result of this appendix (Theorem 3.4), we will prove is:

THEOREM C.3. *Let M be a compact radiant affine 3-manifold with a nonempty totally geodesic boundary. Suppose that each component is affinely homeomorphic to the quotient of a convex cone or* $\mathbf{R}^2 - \{O\}$ *by an affine action. Then M admits a total cross-section to the radial flow, and hence is a generalized affine suspension.*

We cannot prove this theorem by a method of "doubling": Some radiant affine 3-manifold N may not be doubled; i.e., there may not be a radiant affine 3-manifold homeomorphic to the topological double of N in which N and a copy of N are affinely imbedded, meeting at boundary. An example comes from a convex real projective surface Σ with negative Euler characteristic and the boundary component with holonomy ϑ that has a nondiagonalizable 3×3-matrix with two distinct positive eigenvalues ([**11**] and [**13**].) These real projective surfaces exist, of course, as one can see that the construction of convex real projective surfaces in Goldman [**23**] can be modified to construct convex ones with this behavior. Such a holonomy ϑ does not commute with a projective automorphism in $\mathbf{R}P^2$ whose fixed points comprise a subspace that ϑ preserves. We can easily see that the Benzécri suspension of Σ cannot be doubled in the above sense.

The proof of the theorem is essentially that of Theorem B in [**3**] where the totally geodesic surface now is in the boundary.

We remark that the theorem is true without the assumption on boundary component which can be proved applying Barbot's method [**3**]. For simplicity of proof, we prove this weaker but sufficient version here.

The proof of the first one goes as follows: we assume the existence of a radiant affine 3-manifold whose developing map is an infinite cyclic covering over $\mathbf{R}^3 \setminus \{x = y = 0\}$ where x, y, and z denote the standard coordinate functions of $\mathbf{R}^3$.

In the first section, we prove that the holonomy group is solvable: indeed, if not, it contains a hyperbolic element $\rho(\gamma)$ with one eigenvalue greater than 1, another less than 1 (and positive), and the last exactly equal to 1. The contradiction nearly arises from the fact that such a linear transformation does not act properly discontinuously on $\mathbf{R}^3 \setminus \{x = y = 0\}$, whereas γ must act properly discontinuously on $\tilde{M}$.

Since the holonomy group is solvable, the affine 3-manifold is a generalized affine suspension ([**4**]). Therefore, a short argument that no projective surface has a developing map which is an infinite cyclic covering over the sphere minus two points completes the proof.

For the second theorem, we will only prove for the cases when the fact that M has nonempty boundary makes any difference from the proof of Theorem B of [**3**].

If the holonomy of the boundary component contains a homothety, i.e., a linear transformation that is a positive multiple of the identity map, then all radial flow orbits are periodic and it follows easily that our manifold has a total cross-section. First, we show that if the boundary surface is not convex as an affine 2-manifold, then our affine manifold is a half-Hopf manifold. Then we look at the holonomy group of the fundamental group of the boundary component, which we may assume is an affine torus, and classify them into six cases as in [**3**]. Only four cases are applicable since the boundary torus is convex. We will show that in each case, our radiant affine manifold M is finitely covered by a torus times an interval. Either M is foliated by tori corresponding to orbits of certain group actions or decomposes into submanifolds which are affinely isomorphic to domains in $\mathbf{R}^3$ quotient out by linear $\mathbf{Z} + \mathbf{Z}$-action. If M is foliated, there is a quick away to show that M is a generalized affine suspension using Carrière's volume argument [**10**]. Otherwise, we show that the third case is a generalized affine suspension and forth cases are impossible and in the remaining two cases, the pieces must be generalized affine suspensions. (This proof is mostly a generalization of that of Theorem B in Barbot [**3**])

1. The nonexistence of certain radiant affine 3-manifolds

Let M be a closed radiant affine 3-manifold. We denote by Φ^t its radial flow. We denote by $p : \tilde{M} \to M$ the universal covering (we don't worry about the choice of a base point). Let Γ be the fundamental group of M; it acts naturally on $\tilde{M}$.

Let $\mathbf{dev} : \tilde{M} \to \mathbf{R}^3$ be the developing map, and $\rho : \Gamma \to \mathrm{GL}(3, \mathbf{R})$ be the holonomy homomorphism. They satisfy:

$$\forall \gamma \in \Gamma \qquad \mathbf{dev} \circ \gamma = \rho(\gamma) \circ \mathbf{dev}.$$

As above, the radial vector field induces radial flows in M and $\tilde{M}$ respectively. The orbits are said to be *rays* and $\mathbf{dev}$ restricted to each ray is a homeomorphism onto rays in $\mathbf{R}^3$ by Lemma C.1; i.e., an open half-line with an endpoint at 0.

LEMMA C.1. *Let G be a Lie group acting on two manifolds X and Y. Let $f : X \to Y$ be a function equivariant for the actions of G. Let x be an element of X such that $f(x)$ is fixed by no element of G. Then, the restriction of f to the G-orbit of x is injective.*

PROOF. For every element g of G we have $f(gx) = gf(x)$. □

We now assume that $\mathbf{dev}$ is an infinite cyclic covering map over $\mathbf{R}^3 \setminus \Delta$, where Δ is a line through the origin O. Our aim is to obtain a contradiction.

Since we want to show that such a M does not exist, we are free to replace M by any finite covering of itself. For example, we can consider only the case where M is oriented, i.e. the case where every element of the holonomy group is of positive determinant.

Since **dev** is well-defined up to composition by a linear transformation, we can assume that Δ is the line $\{x = y = 0\}$. Then, since Δ has to be $\rho(\Gamma)$-invariant, every element $\rho(\gamma)$ of the holonomy group is of the form:

$$\rho(\gamma) = \begin{pmatrix} \multicolumn{2}{c}{\bar{\rho}(\gamma)} & \begin{matrix} 0 \\ 0 \end{matrix} \\ * & * & \lambda(\gamma) \end{pmatrix}$$

where $\lambda(\gamma)$ is a non-zero real number, and $\bar{\rho}(\gamma)$ an element of $\mathrm{GL}(2, \mathbf{R})$. Clearly, λ and $\bar{\rho}$ are homomorphisms.

We discuss more on generalized affine suspensions (see also Chapter 3): Let $\varphi : N \to N$ a projective diffeomorphism of a real-projective $(n-1)$-manifold N. Recall that $\mathbf{S}^{n-1}$ has a real projective structure induced from $\mathbf{R}P^{n-1}$ by the standard double cover, and the group $\mathrm{Aut}(\mathbf{S}^{n-1})$ of projective automorphisms of $\mathbf{S}^{n-1}$, which is isomorphic to the quotient group of $\mathrm{GL}(n, \mathbf{R})$ by homotheties. ($\mathbf{S}^{n-1}$ with this structure is said to be a *real projective sphere.*) We can always lift the chart of N to $\mathbf{R}P^{n-1}$ to $\mathbf{S}^{n-1}$ with respect to the standard double cover. Then the transition functions then lie in $\mathrm{Aut}(\mathbf{S}^{n-1})$ since a projective map defined on a small open subset of $\mathbf{S}^{n-1}$ extends to one defined on $\mathbf{S}^{n-1}$ always (see Chapter 2 of [**15**]). We gather that N has a natural $(\mathbf{S}^{n-1}, \mathrm{Aut}(\mathbf{S}^{n-1}))$-structure.

Let $f_i : V_i \to U_i \subset \mathbf{S}^{n-1}$ be a family of projective charts covering N. When V_i meets V_j, we have an element $\bar{g}_{ij}$ of $\mathrm{Aut}(\mathbf{S}^{n-1})$ such that on $V_i \cap V_j$:

$$f_i = \bar{g}_{ij} \circ f_j.$$

Let's choose representatives g_{ij} of the $\bar{g}_{ij}$ in $\mathrm{GL}(n, \mathbf{R})$. We impose the condition $g_{ij}g_{jk}g_{ki} = \mathrm{I}$ if $V_i \cap V_j \cap V_k$ is not empty. Such a choice is always possible: take for example the unique representative of $\bar{g}_{ij}$ with determinant ± 1. The set of the possible choices is parameterized by the cohomology group $H^1(N, \mathbf{R})$ of N. For every i, let W_i be the open cone in $\mathbf{R}^n$ with vertex at 0, the union of the half lines belonging to U_i. Let denote by W the quotient of the disjoint union of the W_i by the relation identifying each element x_j of W_j with the element $g_{ij}(x_j)$ of W_i (when $g_{ij}(x_j)$ belongs effectively to W_i, of course). This quotient is a noncompact radiant affine manifold, equipped with a complete radial flow $\hat{\Phi}^t$. The quotient of W by the relation 'being on the same orbit of $\hat{\Phi}^t$' is homeomorphic to N. The quotient map is a fibration by rays. Let N_0 be any section of this fibration. The manifold W is diffeomorphic to $N \times \mathbf{R}$.

Remember the projective transformation φ of N. It lifts to an affine diffeomorphism $\hat{\varphi}$ of W well-defined up to composition by $\hat{\Phi}^t$. If T is big enough, $\hat{\Phi}^T\hat{\varphi}(N_0)$ is a section of $\hat{\Phi}^t$ disjoint from N_0. Therefore, for every real positive t, $\langle \hat{\Phi}^t\hat{\varphi} \rangle$ acts freely and properly discontinuously on W. The quotient of this action is a closed radiant affine manifold homeomorphic to the topological suspension N_φ of $\varphi : N \to N$.

Actually, such a lifting does not always exist for any choice of W, but for many of W above the given Σ, we can perform such liftings. The condition is: let $\bar{\rho} : \pi_1(W) \to \mathrm{GL}(n, \mathbf{R})$ be the holonomy homomorphism of W. Observe that

$\pi_1(W)$ is isomorphic to $\pi_1(N)$. Let φ_* be the automorphism of $\pi_1(N)$ induced by φ. Then, φ lifts if and only if $\det \circ \bar{\rho}$ is constant on the orbits of φ_*. For example, the choice of the g_{ij}'s of determinant ± 1 works.

Observe also that the construction is not uniquely defined: we made some choices. These choices are parameterized by an open subset of the first cohomology module $H^1(N_\varphi, R)$ satisfying the above requirement.

By construction, the radial flow of a generalized affine suspension admits a closed total cross-section homeomorphic to Σ. Note that this section, equipped with the projective structure induced by the transverse projective structure of the radial flow is isomorphic to the initial projective surface Σ. (See Proposition 3.2.)

Returning back to our radiant affine 3-manifold M:

PROPOSITION C.1. *The holonomy group $\rho(\Gamma)$ is solvable.*

PROOF. Denote by Γ' the first commutator subgroup of Γ. Since λ and $\bar{\rho}$ are homomorphisms, for every element of Γ' we have:

- $\bar{\rho}(\gamma)$ belongs to $\mathrm{SL}(2, \mathbf{R})$,
- $\lambda(\gamma) = 1$.

Observe that by definition $\rho(\Gamma)$ is solvable if and only if $\bar{\rho}(\Gamma')$ is solvable.

Let $\mathcal{F}^0$ be the foliation of $\mathbf{R}^3 \setminus \Delta$ whose leaves are the half-planes containing Δ in their boundaries. The leaf space of this foliation, i.e. the quotient of $\mathbf{R}^3 \setminus \Delta$ by the relation "being on the same leaf of $\mathcal{F}^0$", is naturally identified with the double covering of the real projective line $\mathbf{R}P^1$. Let $\mathcal{F}$ be the pull-back of $\mathcal{F}^0$ by $\mathcal{D}$. Since **dev** is an infinite cyclic covering, $\mathcal{F}$ is a foliation whose leaf space is naturally identified with the universal covering $\tilde{P}^1$ of $\mathbf{R}P^1$. The action of Γ' on the leaf space induced by the action of Γ on $\tilde{M}$ is a lifting of the projective action of $\bar{\rho}(\gamma') \in \mathrm{SL}(2, \mathbf{R})$ over $\mathbf{R}P^1$. According to Lemma C.2 below, if $\bar{\rho}(\Gamma')$ is not solvable, there is an element γ of Γ' preserving a leaf F of $\mathcal{F}$, and such that in an adequate coordinate system:

$$\rho(\gamma) = \begin{pmatrix} \lambda & 0 & 0 \\ 0 & \lambda^{-1} & 0 \\ 0 & 0 & 1 \end{pmatrix}$$

for some real positive λ different from 1.

We fix this coordinate system with coordinate functions denoted by x, y, and z still. Since the required coordinate change sends the standard coordinate vectors to the eigenvectors of $\rho(\gamma)$, and the z-axis is in the eigendirection, Δ is still given by $x = 0$ and $y = 0$. Leaves of $\mathcal{F}^0$ are again given as zero sets of linear functions of the new coordinate functions x and y only. We may assume without loss of generality that F maps in the plane given by $x = 0$ under **dev**.

Let P be the inverse image by **dev** of the punctured plane $\{z = 0\} \setminus \{(0, 0, 0)\}$. Since **dev** is an infinite cyclic covering missing Δ, it follows that P is connected, and the restriction of **dev** to P is an infinite cyclic covering to the punctured plane. Moreover, γ preserves P. Since γ preserves the leaf F also, γ preserves each connected component of $\mathbf{dev}^{-1}(\{z = x = 0\})$ and, by same reason, $\mathbf{dev}^{-1}(\{z = y = 0\})$. Let C be a connected component of $\mathbf{dev}^{-1}(\{z = 0, x \geq 0, y \geq 0\})$. It is preserved by γ, and the restriction of **dev** to C is a homeomorphism over $\{z = 0, x \geq 0, y \geq 0\} \setminus \{(0, 0, 0)\}$. The action of $\rho(\gamma)$ on $\{z = 0, x \geq 0, y \geq 0\} \setminus \{(0, 0, 0)$ is given by $(x, y, 0) \mapsto (\lambda x, \lambda^{-1} y, 0)$. It is not properly discontinuous, since any path joining $\{z = 0, x = 0, y \geq 0\}$ to $\{z = 0, x \geq 0, y = 0\}$ intersects all its iterates by $\rho(\gamma)$. This is a contradiction since the action of γ on C has to be properly discontinuous. It follows that $\bar{\rho}(\Gamma)$, and therefore $\rho(\Gamma)$, is solvable. □

For the proof of the following lemma C.2, we must first recall some facts about the action of $\mathrm{PSL}(2, \mathbf{R})$ on $\mathbf{R}P^1$, and the universal covering group $\widetilde{\mathrm{SL}}(2, \mathbf{R})$ action on the universal covering space $\tilde{P}^1$ of $\mathbf{R}P^1$. Let $q : \widetilde{\mathrm{SL}}(2, \mathbf{R}) \to \mathrm{PSL}(2, \mathbf{R})$ denote the covering map. Every element g of $\mathrm{PSL}(2, \mathbf{R})$ is either:

- *elliptic:* g has no fixed point in $\mathbf{R}P^1$. It is conjugate to a rotation,
- *parabolic:* g has one and only one fixed point. This fixed point is of saddle-node type, i.e. attractive on one side, and repulsive on the other side,
- *hyperbolic:* g has two fixed points: a repulsive one and an attractive one. It is conjugate to the element represented by:

$$\begin{pmatrix} \lambda & 0 \\ 0 & \lambda^{-1} \end{pmatrix}.$$

An element of $\widetilde{\mathrm{SL}}(2, \mathbf{R})$ is said to be *elliptic, parabolic or hyperbolic* according to the nature of its projection $q(g)$. If this projection is trivial, g belongs to the center H of $\widetilde{\mathrm{SL}}(2, \mathbf{R})$. The group H is infinite cyclic. Let h be a generator of H. If g is not trivial and admits fixed points on the universal covering $\tilde{P}^1$ of $\mathbf{R}P^1$, it is parabolic or hyperbolic. In the first case, its fixed points are of saddle-node type; in the second case, they are attractive or repulsive.

LEMMA C.2. *Let Γ be a subgroup of $\widetilde{\mathrm{SL}}(2, \mathbf{R})$. We assume that $q(\Gamma)$ is not solvable. Then, it contains a hyperbolic element that fixes a point of $\tilde{P}^1$.*

PROOF. Note that Γ is not solvable since it is a cyclic extension of $q(\Gamma)$ which is not solvable. According to Hölder's theorem (see e.g. [**25**], IV.3.1), a group acting freely on the real line is abelian. Therefore, the action of Γ on $\tilde{P}^1$ is not free: some element γ_0 of Γ admits a fixed point x_0 in $\tilde{P}^1$. If γ_0 is hyperbolic, we are done. If not, γ_0 is parabolic. Then, the fixed points of γ_0 are the H-iterates of x_0. We denote by x_i the image of x_0 by h^i. Observe that since $\tilde{P}^1$ is homeomorphic to the real line $\mathbf{R}$, orienting $\tilde{P}^1$ is equivalent to equip it with an archimedian total order. We orient $\tilde{P}^1$ in such a way that x_1 is greater than x_0. Taking the inverse of γ_0 if necessary, we can assume that all the γ_0-orbits in the open interval $]x_0, x_1[$ go from x_0 to x_1.

The stabilizer in $\mathrm{PSL}(2, \mathbf{R})$ of a point in $\mathbf{R}P^1$ is isomorphic to the group of affine transformations of the line. It is therefore solvable. It follows that there is an element γ of Γ such that $\gamma(x_0)$ is not one of the x_i's. Let γ_1 be the conjugate $\gamma\gamma_0\gamma^{-1}$. It is parabolic and fixes $\gamma(x_0)$. Therefore, it admits a fixed point x'_0 in $]x_0, x_1[$. Then, $\gamma_1^{-1}\gamma_0(x_0) = \gamma_1^{-1}(x_0)$ is less than x_0, and $\gamma_1^{-1}\gamma_0(x'_0)$ is greater than x'_0, since x'_0 is a fixed point of γ_1^{-1} and $\gamma_0(x'_0)$ is greater than x'_0. Therefore, the closed interval $[x_0, x'_0]$ is contained in its image by $\gamma_1^{-1}\gamma_0$. It follows that $\gamma_1^{-1}\gamma_0$ is a hyperbolic element admitting a repulsive fixed point in $]x_0, x'_0[$. □

We know from Proposition C.1 that the holonomy group is solvable. It follows from Theorem A of [**4**] that M is affinely isomorphic to a generalized affine suspension. In particular, the radial flow admits a total cross-section.

Let Σ be such a total cross-section, and $\tilde{\Sigma}$ a lifting of Σ in $\tilde{M}$, i.e. a connected component of $p^{-1}(\Sigma)$. Let $\tilde{\Phi}^t$ be the lifting of Φ^t in $\tilde{M}$. Since Σ is a total cross-section, it is a fiber of some fibration of M over the circle. Hence, $\tilde{M} \setminus \tilde{\Sigma}$ is not connected. Every orbit of $\tilde{\Phi}^t$ meets $\tilde{\Sigma}$. This orbit remains in the past in one connected component of $\tilde{M} \setminus \tilde{\Sigma}$, and in the future, it remains in the other connected component. In other words, every orbit of $\tilde{\Phi}^t$ meets $\tilde{\Sigma}$ at one and only one point. The developing map sends injectively every orbit of $\tilde{\Phi}^t$ over a half-line in $\mathbf{R}^3$ (Lemma C.1 applied to the $\mathbf{R}$-actions). Therefore, it induces a local homeomorphism $\mathbf{dev}'$ from $\tilde{\Sigma}$ on the sphere $\mathbf{S}^2$ of half-lines. We denote by $\mathbf{S}_*$ the sphere $\mathbf{S}^2$ punctured at $(0,0,1)$ and $(0,0,-1)$. Since $\mathbf{dev}$ is an infinite cyclic covering over $\mathbf{R}^3 \setminus \Delta$, the map $\mathbf{dev}'$ is an infinite cyclic covering over S_*. Therefore, $\tilde{\Sigma}$ is the universal covering of Σ, and $\mathbf{dev}'$ is the developing map of a real projective structure on Σ. The holonomy homomorphism $\hat{\rho}$ of this structure is the composition of the restriction of ρ to $\hat{\Gamma}$ with the projection of $\mathrm{GL}(3,\mathbf{R})$ in $\mathrm{Aut}(\mathbf{S}^2)$, where $\hat{\Gamma}$ is the group of elements of Γ which preserve $\tilde{\Sigma}$.

In order to find a contradiction, i.e., in order to achieve the proof of Theorem C.1, it suffices to show:

PROPOSITION C.2. *Given a real projective structure on the closed surface* Σ, *its developing map* $\mathbf{dev}'$ *can not be an infinite cyclic covering over* $\mathbf{S}_*$.

PROOF. Suppose that Σ is a closed surface with such a structure. We first complete $\tilde{\Sigma}$ by the path-metric induced from the Riemannian metric μ on $\mathbf{S}^2$ by $\mathbf{dev}'$ obtaining the Kuiper completion $\check{\Sigma}$ of Σ. Recall that $\tilde{\Sigma}_\infty$ denotes the set of ideal points $\check{\Sigma} - \tilde{\Sigma}$. The developing map $\mathbf{dev}'$ also extends to an obvious distance decreasing map : $\check{\Sigma} \to \mathbf{S}^2$. In our situation, it is easy to see that there exist only two ideal points in $\check{\Sigma}$, mapping to $(0,0,1)$ and $(0,0,-1)$, and that $\check{\Sigma}$ is obtained from $\tilde{\Sigma}$ by adding these two points.

Our surface Σ is obviously not convex since $\tilde{\Sigma}$ is not convex. A 2-*crescent* in $\check{M}$ is a convex hemisphere or lune D in $\check{M}$ with interior in $\tilde{M}$ and the interior of a convex segment in the boundary δD of D includes the nonempty $\tilde{M} \cap \delta D$. Theorems 4.6 and 4.5 of [**15**] show that $\check{\Sigma}$ includes a 2-crescent (see also Section 5 of [**11**]). By definition of 2-crescents, there exists a nontrivial open arc in the boundary of the 2-crescents that is in $\tilde{\Sigma}_\infty$, and hence the set of ideal points in a 2-crescent is uncountable. However, $\check{\Sigma}$ contains only two ideal points. This is a contradiction. □

2. Radiant affine 3-manifolds with boundary have total cross-sections

Now we begin the proof of Theorem C.3. Let M be a compact radiant affine 3-manifold with totally geodesic boundary. Since we may prove the result for a finite cover of M, we assume without loss of generality that the boundary components of M are tori.

LEMMA C.3. *A radiant affine* 3-*manifold* M *admits a total cross-section if and only if it has a closed* 1-*form taking a positive value for each radiant vector.*

PROOF. The existence of a total cross-section and the flow is transversal to it shows that M is diffeomorphic to a bundle over a circle so that the radiant vector field corresponds to the vector field transversal to each fiber. The differential of the fiber map gives us the closed form.

Given a closed form with above property, we can approximate it by a non-vanishing closed form with rational period. Such a closed form obviously gives a fibration $M \to \mathbf{S}^1$ (see [**46**]). □

LEMMA C.4. *If M is a radiant affine 3-manifold, and a finite cover N of M admits a total cross-section to the radial flow, then M admits a total cross-section.*

PROOF. A total cross-section in N corresponds to a closed 1-form on N which is positive for radial vectors. Clearly such a 1-form descends to one on M by averaging over the finite group action as the action preserves the flow direction. $\square$

Let $\tilde{M}$ be the universal cover of M with a development pair $(\mathbf{dev}, h)$. The radiant flow lines induce a *radiant foliation* on $\tilde{M}$. Let Q be the space of leaves of the radiant foliation in $\tilde{M}$, which has a natural real projective structure. The group of deck transformations acts on Q as a group of projective automorphisms (see Barbot [3] for details). As $\tilde{M}$ is simply connected, Q is simply-connected also.

There is a quotient map $f : \tilde{M} \to Q$ which is a fibration whose fibers are rays. We see that the developing map $\mathbf{dev}$ induces an immersion $\mathbf{dev}' : Q \to \mathbf{S}^2$ where $\mathbf{S}^2$ is the space of rays in $\mathbf{R}^3$, and the deck-transformation group acts on Q so that $h'(\vartheta) \circ \mathbf{dev}' = \mathbf{dev}' \circ \vartheta$ for a deck transformation ϑ and $h'(\vartheta)$ the induced projective map from $h(\vartheta)$.

Choose a boundary component K of M and a component $\tilde{K}$ of $p^{-1}(K)$. As K is tangent to the radial flow, K has Euler characteristic zero. By taking a finite cover of M, K is assumed to be a torus. A deck-transformation group G which is a subgroup of $\pi_1(M)$ and isomorphic to $\mathbf{Z}$ or $\mathbf{Z}+\mathbf{Z}$, acts on $\tilde{K}$. Let c be the image of $\tilde{K}$ in Q, which is a simple geodesic in the boundary ∂Q of Q.

An affine automorphism $\varphi : \tilde{M} \to \tilde{M}$ is a *homothety* if each ray is preserved and $\mathbf{dev} \circ \varphi = s\mathrm{I} \circ \mathbf{dev}$ for a positive scalar s. Note that an affine automorphism φ of $\tilde{M}$ always induces a real projective automorphism of Q and φ acts trivially on Q if and only if φ is a homothety.

Let ϑ be an element of G. Then ϑ is not a homothety. If not, then the radial flow is periodic, and M is easily shown to be a Benzécri suspension by Proposition 3.3 of [3]. (Note that for these arguments, there is no difference when M is closed or has nonempty boundary.) We are done in this case.

As $\tilde{K}$ is totally geodesic, $c = f(\tilde{K})$ is a geodesic boundary component of Q. Suppose that K is compressible in M. Then $\tilde{K}$ is compressible in $\tilde{M}$. Let D be an imbedded compressing disk with boundary in $\tilde{M}$. Then the radial projection $f|D : D \to Q$ maps a disk D onto Q with boundary ∂D onto a boundary component $c = f(\tilde{K})$ of Q. This means that Q is homeomorphic to a compact surface, and hence Q is a compact disk. As Q is bounded by a geodesic c, Q must be projectively diffeomorphic to a 2-hemisphere, and $\mathbf{dev}'$ is an embedding onto a 2-hemisphere in $\mathbf{S}^2$. (This can be seen by a doubling argument and the uniqueness of the projective structure on $\mathbf{S}^2$.) This shows that $\tilde{M}$ is affinely diffeomorphic by $\mathbf{dev}$ to an affine half-space with boundary containing O with O removed.

PROPOSITION C.3. *Suppose that Q is real projectively diffeomorphic to a hemisphere in $\mathbf{S}^2$, or equivalently $\tilde{M}$ is affinely diffeomorphic to an affine half-space H_1 with boundary containing O with O removed. Then M is a half-Hopf manifold.*

PROOF. We will use the second hypothesis. The punctured half-plane $H_1 - \{O\}$ includes a compact disk D with boundary in $\partial H_1 - \{O\}$ which is transversal to every ray. There exists a deck transformation ϑ of $H_1 - \{O\}$ sending D to $\vartheta(D)$ disjoint from D as the deck-transformation groups are properly discontinuous. Clearly $H_1 - \{O\}$ quotient out by ϑ has a total cross-section corresponding to D, and is a half-Hopf manifold. Hence, M is finitely covered by a generalized affine suspension, and so M is a generalized affine suspension. The total cross-section has positive Euler characteristic, and hence is a compact disk transversal to radial flow. The lemma now follows. □

Assume that K is incompressible from now on. First, suppose that K is affinely homeomorphic to a quotient of $\mathbf{R}^2 - \{O\}$ by an affine action. Now, we need to use the *holonomy cover* M_h of M, i.e., the cover of M corresponding to the kernel of the holonomy homomorphism h. As we described in [**15**], the developing map **dev** induces an immersion $\mathbf{dev}_h : M_h \to \mathbf{R}^3$ and the holonomy homomorphism induces a homomorphism $h_h : \pi_1(M)/\pi_1(M_h) \to \mathrm{GL}(3, \mathbf{R})$. Also f induces a fibration $f_h : M_h \to Q_h$ to a projective surface Q_h covered by Q. The immersion $\mathbf{dev}_h$ induces an immersion $\mathbf{dev}'_h : Q_h \to \mathbf{S}^2$ and h_h induces a homomorphism $h'_h : \pi_1(M)/\pi_1(M_h) \to \mathrm{Aut}(\mathbf{S}^2)$. The surface $\tilde{K}$ corresponds to a surface K_h covering K. We see easily that K_h is affinely diffeomorphic to $\mathbf{R}^2 - \{O\}$. A deck-transformation group G_h acts on K_h so that K is affinely homeomorphic to K_h/G_h. Hence G_h is an infinite cyclic group.

A *homothety* in M_h is an affine automorphism φ of M_h acting on each ray and satisfying $\mathbf{dev}_h \circ \varphi = s\mathrm{I} \circ \mathbf{dev}_h$ for a positive scalar s. As above, if an element of G_h is a homothety, then M is a generalized affine suspension.

Assume now that no element of G_h is a homothety. We claim that in this case M is a half-Hopf manifold. Let ϑ be a generator of G_h. Then $h_h(\vartheta)$ acts on the totally geodesic plane P including $\mathbf{dev}_h(K_h)$. We assume that P is the xy-plane for simplicity. As no element of G_h is a homothety, G_h acts effectively on Q_h. We divide our cases according to the conjugacy classes of $h'_h(\vartheta)$ in $\mathrm{Aut}(\mathbf{S}^2)$. Let P' be the great circle in $\mathbf{S}^2$ corresponding to P, and H_1 and H_2 the two hemispheres bounded by P'. We easily see that one of the following occurs:

(i) The only $h'_h(\vartheta)$-invariant subset of H_i including a neighborhood of P' in H_i is H_i itself.
(ii) The only subset of H_i with this property equals $H_i - \{x\}$ for a point $x \in H_i^o$.
(iii) $h'_h(\vartheta)$ under a suitable coordinates of the affine space H_i^o is of form

$$\begin{pmatrix} \cos 2\pi\theta & \sin 2\pi\theta \\ -\sin 2\pi\theta & \cos 2\pi\theta \end{pmatrix}, \theta \neq 0.$$

Note that $\mathbf{dev}'_h|c$ is an imbedding onto P', and hence a neighborhood U of c in Q_h imbeds onto a neighborhood of P' in say H_1. We see that $\mathbf{dev}'_h \circ \vartheta|U = h'_h(\vartheta) \circ \mathbf{dev}'_h|U$, and so $\mathbf{dev}'_h|\vartheta(U)$ is also an imbedding onto a neighborhood of P'. Therefore, by induction, we see that $\mathbf{dev}'_h|\bigcup_{i\in\mathbf{Z}} \vartheta^i(U)$ is an imbedding onto an $h'_h(\vartheta)$-invariant subset of H_1 including a neighborhood of P'.

In case (i), Q_h maps homeomorphic onto H_1, and by Proposition C.3, M is a half-Hopf manifold.

In case (ii), Q_h maps homeomorphic to $H_1 - \{x\}$. Therefore M_h under $\mathbf{dev}_h$ maps homeomorphic to $H'_1 - l$ for the upper half-space H'_1 corresponding to H_1 and a line l through O transversal to the xy-plane. Hence, we may identify M_h with its image, and each deck transformation acts on l and the upper-half space. We see that the group of deck transformations acting on K_h equals the entire deck-transformation group of of M_h and hence the group of deck transformations acting on $\tilde{K}$ equals $\pi_1(M)$. This means that K is homotopy equivalent to M by the inclusion map where $K = \partial M$. We see easily by a topological doubling argument and the top dimensional $\mathbf{Z}_2$-homology group computation that such a situation cannot happen.

In case (iii), θ must be irrational since otherwise $h'_h(\vartheta)$ is periodic and hence ϑ must be periodic near c_h and hence on Q_h; but this means that a power of ϑ is a homothety on M_h.

Under the suitable coordinates, for some large r, there is an open neighborhood U_r of c which under $\mathbf{dev}'_h$ maps homeomorphic to the set of form the union of P' and the complement of a ball of radius r.

Let r_0 be an infimum of possible values of r. Then we see easily that U_{r_0} exists. We claim that (1) Q_h equals this set U_{r_0} or (2) Q_h maps homeomorphic to H_1 under $\mathbf{dev}'_h$. Suppose that there exists a boundary point x of U_{r_0} in Q_h. As ϑ acts as an irrational rotation, we see that every point of the boundary of $\mathbf{dev}'(U_{r_0})$ is an image of a boundary point of U_{r_0} in Q_h. Assume that $r_0 > 0$. In this case, the boundary γ of U_{r_0} in Q_h maps homeomorphic to the closure γ' of the orbit of ϑ in H_1, which is homeomorphic to a circle. If there exists a point of ∂Q_h in the boundary γ, then every point of γ is in ∂Q_h as ∂Q_h is closed and ϑ acts on ∂Q_h. But this implies that a boundary component of M_h is not totally geodesic. Therefore, there exists a regular neighborhood of γ mapping to a regular neighborhood of γ'. This contradicts the minimality of r_0 if $r_0 > 0$. Hence, we obtain that $U_{r_0} = Q_h$. When $r_0 = 0$, the claim follows easily.

In case (1), we obtain that M_h maps homeomorphic under $\mathbf{dev}_h$ to the complement L of a closed convex cone in H'_1 not meeting the boundary $P - \{O\}$. We identify M_h with this set L. Then a contradiction as in (ii) occurs.

In case (2), we see that M is a half-Hopf manifold by Proposition C.3.

From now on, we will be working on $\tilde{M}$ (i.e., not on M_h) and assume that $\tilde{K}$ is affinely homeomorphic to a convex cone in $\mathbf{R}^2$.

Since $h(G)$ acts on a convex cone $\mathbf{dev}(\tilde{K})$, if $h(\varphi)$ for $\varphi \in G$ is a homothety, then φ is a homothety near $\tilde{K}$ in $\tilde{M}$, and hence φ is a homothety on $\tilde{M}$. Thus $q \circ h$ is injective. We identify G with $h(G)$ from now on.

Let $q : \mathrm{GL}(3, \mathbf{R}) \to \mathrm{SL}(3, \mathbf{R})$ be the homomorphism whose kernel consists of $s\mathrm{I}$ for $s \neq 0$. Since no element of G is a homothety in $\tilde{M}$, Barbot [**3**] shows that the identity component L of the Zariski closure of $q(G)$ is conjugate to a subgroup of the following groups:

Case D: the group of all diagonal matrices with positive eigenvalues.

Case P: the group of matrices of form:

$$\begin{pmatrix} e^u & t & 0 \\ 0 & e^u & 0 \\ 0 & 0 & e^v \end{pmatrix}, u, v, t \in \mathbf{R}, 2u + v = 0,$$

Case U: the group of matrices of form:

$$\begin{pmatrix} 1 & s & t \\ 0 & 1 & 0 \\ 0 & 0 & 1 \end{pmatrix}, s, t \in \mathbf{R},$$

Case C: the group of matrices of form:

$$\begin{pmatrix} 1 & s & t \\ 0 & 1 & s \\ 0 & 0 & 1 \end{pmatrix}, s, t \in \mathbf{R},$$

Case S: the group of matrices of form:

$$\begin{pmatrix} e^u \cos\theta & e^u \sin\theta & 0 \\ -e^u \sin\theta & e^u \cos\theta & 0 \\ 0 & 0 & e^v \end{pmatrix}, u, v, \theta \in \mathbf{R}, 2u + v = 0,$$

or

Case T: the group of matrices of form:

$$\begin{pmatrix} 1 & 0 & s \\ 0 & 1 & t \\ 0 & 0 & 1 \end{pmatrix}, s, t \in \mathbf{R}.$$

Since $q(G) \cap L$ must be a finite index subgroup of $q(G)$, we assume that $q(G)$ is a rank-two abelian subgroup of L by taking a finite cover of M and choosing $(\mathbf{dev}, h)$ carefully, i.e., conjugating h by an element of $\mathrm{GL}(3, \mathbf{R})$.

In cases S and T, if L is one-dimensional, then L can be considered a subgroup of D and C respectively by conjugations. In case S, since ϑ acts on $\mathbf{dev}(\tilde{K})$ a convex two-dimensional subset, θ is always 0 for each element of the group G and we are reduced to the one-dimensional case. In case T, assuming that L is two-dimensional, G acts on the xy-plane as a homothety, and there are no other G-invariant subspaces of codimension-one. But since $\mathbf{dev}(K)$ is a G-invariant subspace so that $\mathbf{dev}(K)/G$ is homeomorphic to the torus K/G, this is a contradiction by the following lemma. Hence the case T does not occur.

LEMMA C.5. *Let D be a convex cone in $\mathbf{R}^2$, and G' an abelian group isomorphic to $\mathbf{Z} + \mathbf{Z}$. If D/G is homeomorphic to a torus, then an element of G' is not a homothety.*

PROOF. If each element of G' is a homothety, then each element of G' acts on each ray in D ending at O. Since $\mathbf{Z} + \mathbf{Z}$ cannot act properly discontinuously and freely on a real line, this is absurd. □

We see that G is a lattice in a connected two-dimensional subgroup H of following group of matrices in $\mathrm{GL}(3, \mathbf{R})$:

Case D: the group of all diagonal matrices of positive eigenvalues.

Case P: the group of matrices of form

$$\begin{pmatrix} e^u & t & 0 \\ 0 & e^u & 0 \\ 0 & 0 & e^v \end{pmatrix}, u, v, t \in \mathbf{R}.$$

Case U: the group of matrices of form:

$$\begin{pmatrix} e^u & s & t \\ 0 & e^u & 0 \\ 0 & 0 & e^u \end{pmatrix}, u, s, t \in \mathbf{R},$$

or

Case C: the group of matrices of form:

$$\begin{pmatrix} e^u & s & t \\ 0 & e^u & s \\ 0 & 0 & e^u \end{pmatrix}, u, s, t \in \mathbf{R}.$$

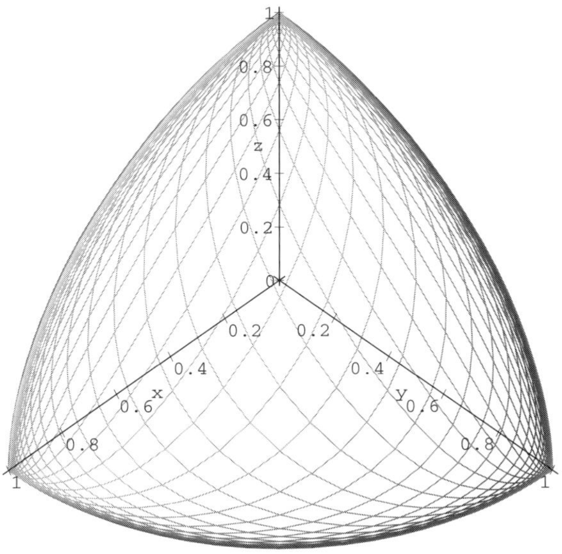

FIGURE C.1. An example of an H-action on $\mathbf{S}^2$ in case D(2). The sphere itself is not drawn, and the arcs correspond to orbits of one-parameter subgroups.

The Lie group $q(H)$ is not zero dimensional, as G does not contain a homothety. $q(H)$ could be one-dimensional or two-dimensional. If $q(H)$ is two-dimensional, then $q(G)$ is a lattice in $q(H)$; otherwise, $q(G)$ is dense in $q(H)$.

The group H acts on the projective sphere $\mathbf{S}^2$ effectively as a subgroup of the group $\mathrm{Aut}(\mathbf{S}^2)$ of projective automorphisms of $\mathbf{S}^2$. The space $\mathbf{S}^2$ decomposes into two-dimensional open orbits and one-dimensional orbits and zero-dimensional orbits under H. We have to divide our case to when $q(H)$ is one-dimensional and when two-dimensional. (1) indicates the one dimensional cases which differ from (2) by having no two-dimensional orbits. The zero dimensional orbits are the same, and there are additional one-dimensional orbits, which foliate the regions corresponding to two-dimensional orbits.

Case D(2): The zero-dimensional orbits are six points in $\mathbf{S}^2$ comprising three pair of antipodal points, one-dimensional ones are twelve lines in $\mathbf{S}^2$ with endpoints in the three points, and two-dimensional orbits are eight open triangles bounded by the closures of the lines. (See Figure C.1.)

Case D(1): The additional one-dimensional orbits are curves curved in one direction connecting a fixed pair of zero-dimensional orbits, or lines that are components of great circles passing through a common antipodal pair of zero-dimensional orbits with the union of the above twelve lines removed.

Case P(2): The zero-dimensional orbits comprise two pair of antipodal points $\{p, -p\}$ and $\{q, -q\}$, one-dimensional orbits are four lines with endpoints $\pm p$

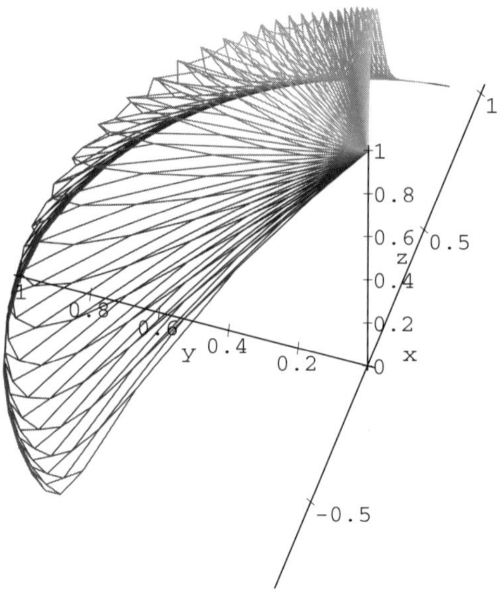

FIGURE C.2.
The case P(2): the orbits are partially drawn here.

and $\pm q$ and two lines that are components of a great circle containing $p, -p$ with p and $-p$ removed, and two-dimensional orbits are four open lunes bounded by the closure of the union of one-dimensional orbits. (See Figure C.2.)

Case P(1): The additional one-dimensional orbits are curves connecting a pair of zero-dimensional orbits curved in one-direction.

Case U(2): Zero-dimensional orbits are two points p and $-p$. One-dimensional orbits are great circles containing p and $-p$ with p and $-p$ removed. Their union equals $\mathbf{S}^2 - \{p, -p\}$ and there are no two-dimensional orbits.

Case U(1): Zero-dimensional orbits are points of a great circle through two points $\{p, -p\}$. One dimensional orbits are the other great circles through p and $-p$ with p and $-p$ removed. There are no two-dimensional orbits.

Case C(2): Zero-dimensional orbits are two points p and $-p$. One-dimensional orbits are two components of a great circle containing p and $-p$ with $\{p, -p\}$ removed. Two-dimensional orbits are two open hemispheres bounded by the great circle. (See Figure C.3.)

Case C(1): The additional one-dimensional orbits are curves in the open hemisphere connecting p to p or $-p$ to $-p$, they are curved in one directions.

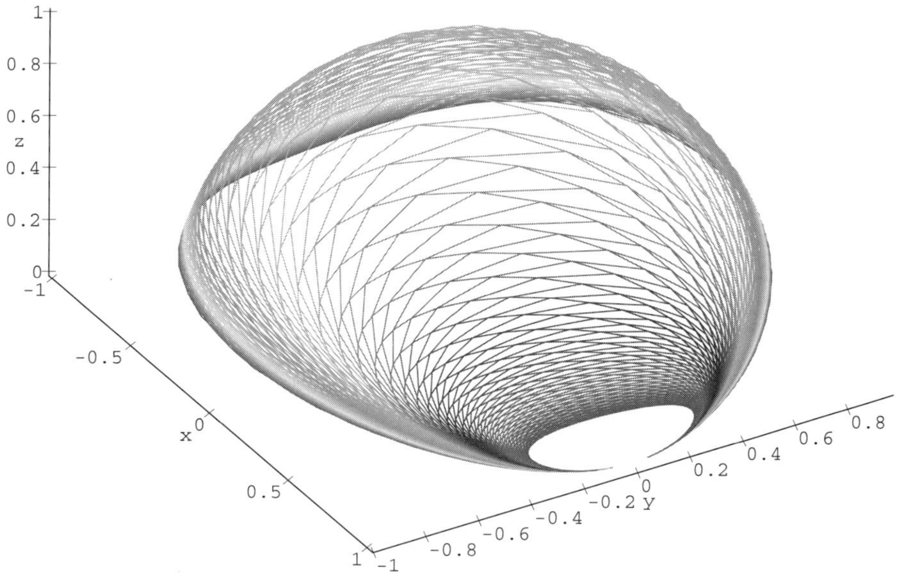

FIGURE C.3. The case C(2).

We see that all one-dimensional orbits are simple geodesics in $\mathbf{S}^2$ and are not closed ones when $q(H)$ is two-dimensional. (To see some more figures of these kinds of actions, see [**13**].)

Recall that the developing map $\mathbf{dev} : \tilde{M} \to \mathbf{R}^3$ induces an immersion $\mathbf{dev}' : Q \to \mathbf{S}^2$ and the deck-transformation group acts on Q as well so that $\mathbf{dev}' \circ \vartheta$ equals $h'(\vartheta) \circ \mathbf{dev}'$ where $h'(\vartheta) = q(h)(\vartheta)$ for each deck transformation ϑ of $\tilde{M}$. (The action is not necessarily proper.)

LEMMA C.6 (Lemma 2.4.3 of Dupont [**18**]). *Let V be a (U, X)-manifold where U is a Lie group acting on a space X. Let L be a connected subgroup of U, and ω an L-orbit in X and $\hat{\omega}$ a connected component of $D^{-1}(\omega)$ where (D, j) is a development pair of V. Suppose that the action of L on ω is covered by the action of L on $\hat{\omega}$, and there exists a subgroup Γ_0 of the deck-transformation group so that $\hat{\omega}/\Gamma_0$ is compact and $j(\Gamma_0)$ is in L. Then the action of L on a neighborhood of ω is covered by an action of L on a neighborhood of $\hat{\omega}$ such that, for any element γ of Γ_0, the action of $j(\gamma)$ coincides with the action of γ as a deck transformation.*

Since G acts on $f(\tilde{K}) \subset Q$ as in the premise of the above lemma, the extension argument (i.e., the proof of Theorem B in [**3**] using essentially Lemma C.6) shows that H acts on Q, Q is a union of orbits of dimension zero, one, or two, and each orbit maps homeomorphic to an orbit in $\mathbf{S}^2$ under $\mathbf{dev}'$. The proof of this claim, similar to what is in [**3**], goes as follows: Let $\mathcal{O}$ be the maximal connected subset of Q including c where the H-action is defined and whose restriction to G coincide with the deck transformation action. Our claim follows from the following lemma:

PROPOSITION C.4. *The H-action is defined everywhere; i.e. $\mathcal{O}$ equals Q.*

PROOF. We have to give a more precise definition of $\mathcal{O}$: this is the maximal connected subset including c of the set of points x in Q such that:

- There is a continuous action of H on Q, denoted by $(k, x) \in H \times Q \mapsto k.x$ so that, for any (k, x), we have $\mathbf{dev}'(k.x) = k\mathbf{dev}'(x)$.
- for every element g of G, we have the equality $g.x = gx$ (remember that G is a group of deck transformations, and therefore acts on Q).

The fundamental group of an orbit of the radial flow is trivial or cyclic. It follows that there are no zero-dimensional H-orbits in $\mathcal{O}$. (An abelian group of rank 2 such as G cannot act on a connected one-dimensional space properly discontinuously and freely.)

By Lemma C.6, $\mathcal{O}$ is open. Since Q is connected, the lemma will be proven if we show that $\mathcal{O}$ is closed.

As H acts on $\mathcal{O}$, $\mathcal{O}$ is a union of one- or two-dimensional H-orbits in Q. The restriction of $\mathbf{dev}'$ to each H-orbit maps homeomorphic to an H-orbit in $\mathbf{S}^2$ by Lemma C.1. In particular, two-dimensional H-orbits in $\mathcal{O}$ are open surfaces, and G acts properly discontinuously and freely on each of them.

We note that each two-dimensional H-orbit in $\mathcal{O}$ has to have at least one adjacent one-dimensional H-orbit. If not, the two-dimensional orbit is disconnected from c, a contradiction.

We see that $\mathcal{O}$ is a union of two-dimensional orbits and adjacent one-dimensional orbits joined in a "chain-like" manner in cases D(2), P(2), and C(2) or is a union of one-dimensional orbits in other cases.

Suppose that x is a boundary point of $\mathcal{O}$ in Q. We aim to obtain a contradiction. If $\mathbf{dev}'(x)$ lies in a zero-dimensional orbit of H, then x is a zero-dimensional orbit of H. This is a contradiction by above. The image $\mathbf{dev}'(x)$ does not lie in a two-dimensional orbit as each two-dimensional H-orbit of Q is open in Q and $\mathcal{O}$ is a union of H-orbits of Q. Let $\mathbf{dev}'(x)$ be in a one-dimensional orbit J, and let J' be a component of $\mathbf{dev}'^{-1}(J)$ containing x. Then J' is a closed subset of Q. The restriction of $\mathbf{dev}'$ to J' is injective.

Suppose that the intervals J and $\mathbf{dev}'(J')$ have a common endpoint. Then, for any element g of G, the intervals $\mathbf{dev}'(J')$ and $g\mathbf{dev}'(J')$ are not disjoint. Actually, inverting g if necessary, we can assume that $\mathbf{dev}'(J')$ contains $g\mathbf{dev}'(J')$. Since the restriction of $\mathbf{dev}'$ to every H-orbit in $\mathcal{O}$ is injective, a union A of some H-orbits near J' is an open set where H acts, and the restriction of $\mathbf{dev}'$ to A is injective. Since $\mathbf{dev}'$ restricted to $A \cup J'$ and $A \cup g(J')$ are both injective, J' contains $g(J')$. It follows that J' is G-invariant. We deduce that J' is contained in $\mathcal{O}$; a contradiction. (*)

Therefore, the endpoints of $\mathbf{dev}(J')$ both belong to J.

Suppose that no element of G fixes a point of J. Then, all the G-orbits in J are dense. There is an element g of G such that $g(\mathbf{dev}'(J'))$ meets $\mathbf{dev}'(J')$. Since as above $\mathbf{dev}'$ restricted on $A \cup J'$ and $A \cup g(J')$ are both injective, J' is g-invariant, meaning that H acts on J' again.

This contradiction shows that some element g of G acts as an identity map on J. Therefore, the surface D in $\tilde{M}$ corresponding to J' is filled with rays on which g acts, and hence, maps to the union of closed orbits of radial flow in M.

We claim that, for any deck transformation ϑ, J' cannot meet $\vartheta(J')$ where J' and $\vartheta(J')$ are distinct. Suppose not. Then D and $\vartheta(D)$ meet at an isolated ray l by the real analyticity of J' and $\vartheta(J')$. This ray is fixed by g; hence, g is a power of a deck transformation g' so that $l/\langle g' \rangle$ maps to the closed orbit in M. Similarly, $\vartheta \circ g \circ \vartheta^{-1}$ is a power of g'. Thus, a finite power g'' of g acts trivially in J' and $g(J')$ since a power of $\vartheta \circ g \circ \vartheta^{-1}$ must equal a power of g. Since we can find four nearby fixed points in Q of g'' contained in J' and $\vartheta(J')$, g'' is a homothety, a contradiction. (To see this, recall that orbits are real analytic submanifolds of Q.)

Also, the collection of sets of form $\vartheta(J')$ are locally finite. If not, then there exists a sequence of points $p_i \in \varphi_i(D)$ converging to $p \in \tilde{M}$ for a sequence of deck transformations φ_i. As $\varphi_i \circ g \circ \varphi_i^{-1}$ acts on $\varphi_i(D)$ as homotheties by a fixed factor $s, s > 0$, we see that $\varphi_i \circ g \circ \varphi_i^{-1}$ moves p_i to a points in a compact subset of $\tilde{M}$. The properness of the deck-transformation-group action shows that infinitely many of these deck transformations must be the same. We may choose $\varphi_i(D)$ and $\varphi_j(D)$ sufficiently close so that $\varphi_i \circ g \circ \varphi_i^{-1}$ equals $\varphi_j \circ g \circ \varphi_j^{-1}$. However this means that $\varphi_i \circ g \circ \varphi_i^{-1}$ is a homothety since it acts trivially on $\varphi_i(J')$ and another curve $\varphi_j(J')$, a contradiction.

By above two conclusions, it follows that D covers a closed surface in M tangent to the radial flow, and a subgroup G' of rank 2 acts on D and on J'. All we did above apply once more: there is a connected abelian group H' including G' as a dense subgroup or a lattice, and H' is again of type D, P, U, or C. (This is easy if J' is geodesic. If not, we look at the line connecting the endpoints of $\mathbf{dev}(J')$.) We define similarly to $\mathcal{O}$ the locus $\mathcal{O}'$ of definition of the H'-action on Q. It contains J'. Since $\mathbf{dev}'(J')$ has both endpoints inside $\mathbf{dev}(J)$, H' cannot be in case U or C, since in these cases, $\mathbf{dev}'(J')$ should equal J' by geometric reasons.

Suppose that H' is in case D(2) or P(2), and H is in case D(2), P(2), or C(2). Then, J' is contained in the boundary of a two-dimensional H-orbit A. By looking at the image under $\mathbf{dev}'$ of A and the two-dimensional H'-orbit B meeting $\mathbf{dev}'(A)$, we obtain that there exists a geodesic c in A mapping into a one-dimensional H'-orbit adjacent to $\mathbf{dev}'(B)$ so that $\mathbf{dev}'(c)$ and $\mathbf{dev}'(J')$ share an endpoint. Since an endpoint of $\mathbf{dev}'(c)$ is an H'-orbit, c must be included in an H'-orbit by the same reason as the paragraph marked by (*) with G replaced by G'. By the following lemma C.8, this is a contradiction.

If H is in case D(1), P(1), C(1) or U(1)(2), then as in the above paragraph a one-dimensional H'-orbit meets a one-dimensional H-orbit transversally. (We can see this by the transversality theorem applied to the foliation by orbits, i.e., a parameter of submanifolds [**24**].) This is a contradiction by Lemma C.7.

Suppose that H' is in case D(1) or P(1), and H is in case D(2), P(2), or C(2). Then clearly a one-dimensional orbit of H' near J' meets a two-dimensional orbit of H adjacent to J', a contradiction by Lemma C.8. As above, H cannot be in case D(1), P(1), C(1) or U(1)(2), This final contradiction achieves the proof of Proposition C.4. □

LEMMA C.7. *Let H' be a connected two-dimensional abelian group including an abelian group G' of rank 2 of deck transformations of $\tilde{M}$. Then a one-dimensional orbit K' of H' in $\mathbf{S}^2$ does not meet a one-dimensional orbit K of H in $\mathbf{S}^2$ transversally.*

PROOF. Suppose that K and K' meet at a point x in Q. Then let S and S' be the corresponding surfaces in $\tilde{M}$. They meet at a ray l corresponding to x. As S/G and S'/G' correspond to immersed tori, l corresponds to a closed orbit of the radial flow in M. There is an element of g in $G \cap G'$ corresponding to this orbit.

Then $\langle g \rangle$ acts trivially on K and K' as we can see from the matrices of form D, P, U, and C; i.e., G acts on one-dimensional orbits in $\mathbf{S}^2$ as translations with respect to certain coordinates. Since we can find four nearby fixed points of g in $K \cup K'$, the holonomy of g must be a homothety. This contradicts our assumption. □

LEMMA C.8. *Let H' and G' be as in the preceding lemma. Let A be a two-dimensional H-orbit including in its boundary a one-dimensional H-orbit C included in $\mathcal{O}$. Then, no one-dimensional H'-orbit included in $\mathcal{O}'$ meets A where $\mathcal{O}'$ is the region of Q where H' action is defined near C.*

PROOF. Since G is a lattice of H, for every element x of A, there is a sequence of elements h_n of G for which the sequence $h_n.x$ converges to some point of C by Lemma C.9. Assume that some one-dimensional H'-orbit B meets A. Then, some iterates $h_n.b$, where b belongs to $B \cap A$, accumulates to a point in C. But B and C correspond to some immersed tori in M; therefore, such an accumulation is impossible. □

LEMMA C.9. *Let H'' be a connected two-dimensional abelian group of projective transformations of the projective sphere $\mathbf{S}^2$. Let G'' be a lattice of H'', and let A be an open orbit of H'' in the projective plane. Let C be a one-dimensional orbit of H'' contained in the boundary of A. Then, the closure of the G''-orbit of any element of A contains a point of C.*

PROOF. Let F be a compact connected fundamental domain for the action of G'' on H'' by translations. Let a be any point of A, and x a point of C. The orbit $F.x$ of x by F is a compact part of C. Let U be an open neighborhood of $F.x$ in the projective plane. By continuity of the action of H'' on the real projective plane, and by compactness of F, there is an open neighborhood V near x such that for every element v of V, the orbit $F.v$ is contained in U. The neighborhood V certainly contains an element of the orbit of a by H''. Since H'' is abelian, and since F is a fundamental domain, $F.V$, and thus U, must contain an element of the orbit of a by G. Since this is true for any open neighborhood U of $F.x$, it follows that the orbit $G''.a$ accumulates at least on some point of $F.x$. □

We claim that the decomposition of Q into H-orbits are preserved under the action of the deck-transformation group. A one-dimensional H-orbit l in Q does not meet with an image $\vartheta(m)$ of a one dimensional orbit m in Q transversally for a deck transformation ϑ by Lemma C.7. A one-dimensional H-orbit does not meet an image of two-dimensional H-orbit under a deck transformation by Lemma C.8. Also, if a one-dimensional orbit l meets an image of $\vartheta(m)$ nontransversally with $\vartheta(m)$ not equal to l, then either a transversal intersection of a one-dimensional orbit with $\vartheta(m)$ or the intersection with two-dimensional orbits must occur nearby. Thus, H-orbits map to H-orbits under deck transformations as there are no zero dimensional H-orbits in Q.

We can quickly achieve the proof of Theorem C.3 when $q(H)$ is one-dimensional. Each $q(H)$-orbit has a G-invariant one-dimensional volume form on it as $q(H)$ is a one-dimensional Lie group with a left-invariant volume form. Note that the volume form depends smoothly on orbits. As G acts on each of the subsets of $\tilde{M}$ corresponding to one-dimensional orbits in Q, these sets cover tori or Klein bottles under the covering map $\tilde{M} \to M$. These tori or Klein bottles fiber M. By taking a finite cover of M if necessary, we can assume that M is homeomorphic to a torus times an interval with fibers corresponding to the tori. Hence, there exists a volume form transverse to the tori which becomes a transverse volume form of the orbits. Therefore, the volume form extends to a smooth two-dimensional volume form on Q. Hence, M has a flow invariant three-dimensional volume form. By Proposition 4 of Carrière [**10**], we see that M admits a total cross-section.

From now on, we assume that $q(H)$ is two-dimensional. We say that an affine 3-manifold M *decomposes* into affine 3-manifolds $N_1, \dots, N_n$ if each N_i is the closure of a component of M with two-sided separating totally geodesic surfaces in M removed.

PROPOSITION C.5. *Our manifold M or a finite cover of M decomposes into compact radiant affine 3-manifolds N_i each of which is affinely isomorphic to a quotient of a domain in $\mathbf{R}^3 - \{O\}$ by an action of G. Every piece N_i is homeomorphic to torus times an interval and has totally geodesic boundary which are tori isotopic to K or a finite cover of K.*

PROOF. The H-orbits in $\mathbf{S}^2$ correspond to H-invariant submanifolds in $\mathbf{R}^3 - \{O\}$. Say these sets are *H-invariant sets* in $\mathbf{R}^3 - \{O\}$. We now choose domains in $\mathbf{R}^3 - \{O\}$ consisting of adjacent three- and two-dimensional H-invariant sets. In cases D, P, C, an *H-domain* is the union of an H-invariant open set and H-invariant two-dimensional sets in the boundary of the set. In case U, we choose two antipodal two-dimensional H-invariant sets so that $\mathbf{dev}(\tilde{K})$ is included in one of them. Their complement is the union of two convex H-invariant open sets. Call these sets *distinguished* two-dimensional H-invariant sets. An *H-domain* in case U is the union of one convex open set and the two two-dimensional H-invariant sets included in its boundary.

We look at a component of the inverse image in Q of one-dimensional orbits under $\mathbf{dev}'$ or distinguished orbits in case U. Then they are one-dimensional H-orbits mapping homeomorphic to the orbits below. A component of the complement of these orbits are either two-dimensional orbits or a union of one-dimensional orbits in case U.

On $\tilde{M}$, these orbits correspond to a decomposition into H-invariant open sets and H-invariant totally geodesic two-dimensional sets mapping homeomorphic to their images in $\mathbf{R}^3$, which are also H-invariant.

Recall that H-invariant two-dimensional sets map (distinguished ones in case U) to imbedded closed surfaces in M under the covering map. By taking a finite cover of M if necessary, we assume that these are two-sided tori in M. Take one, say A, of the open H-invariant sets in M. Then since the deck-transformation group acts on $\tilde{M}$ preserving the decomposition, it follows that A union with adjacent two-dimensional orbits map to a closed submanifold in M bounded by tori. Let us denote them by $N_1, \dots, N_h$. Then clearly, M decomposes into $N_1, \dots N_n$.

The manifolds N_i are obviously affinely homeomorphic to the quotients of three-dimensional closed domains which are unions of H-invariant sets.

Take the H-invariant open set A adjacent to $\tilde{K}$, a two-dimensional H-invariant set. Then the closure of A is the union of A and adjacent one, two, or three two-dimensional H-invariant sets. The closure covers a compact radiant affine manifold N_1, i.e, it may be identified with a universal cover $\tilde{N}_1$ of N_1. The deck-transformation group of M acting on A can be identified with the deck-transformation group of $\tilde{N}_1$. Then since the deck-transformation group acts on

A nontrivially, it acts on $\tilde{K}$ also. (Remember that G is a subgroup of H.) It follows that $\pi_1(K) \to \pi_1(N_1)$ is an isomorphism. By three-manifold topology, we may assume that N_1 is homeomorphic to a torus times an interval.

Removing N_1 from M and taking the closure in M, we get a new radiant affine 3-manifold which decomposes into $N_2, \ldots, N_n$. By induction, we see that each N_i is homeomorphic to a torus times an interval. The proof of Proposition C.5 is complete. □

COROLLARY C.1. *Let $p : \tilde{M} \to M$ be the universal covering. Then $p^{-1}(N_i)$ is connected and under p maps as a universal covering map onto N_i. Each $p^{-1}(N_i)$ is a three-dimensional H-invariant domain, and maps homeomorphic to a three-dimensional H-invariant set union with two adjacent two-dimensional H-invariant sets under* **dev**.

We denote by $\tilde{N}_i$ the set $p^{-1}(N_i)$ for simplicity.

From now on, we assume that M satisfies the conclusion of Proposition C.5. We end this section by showing that M must be a generalized affine suspension in the cases D, P, U, and C.

In case U, choose the open H-invariant set $\tilde{A}$ adjacent to $\tilde{K}$. Then $\tilde{A}$ is bounded by $\tilde{K}$ and another orbit L so that the angle θ between $\tilde{K}$ and L is less than or equal to π.

If $\theta < \pi$, then L is another boundary component of $\tilde{M}$, since if not we could have enlarged $\tilde{A}$. Thus, $N_1 = M$ and M is a generalized affine suspension since we can use the 1-form dz/z for a linear function z to get a total cross-section where the plane $z = 0$ is disjoint from the image under **dev**.

Clearly, G acts on $\mathbf{dev}(A) \cup \mathbf{dev}(\tilde{K}) \cup \mathbf{dev}(L)$ properly discontinuously. If $\theta = \pi$, then this set is in the form of a half-space with a line in its boundary removed. By Lemma C.10, this is a contradiction.

In case C, let A be the open H-invariant set adjacent to $\tilde{K}$. The closure of A in $\tilde{M}$ equals $A \cup \tilde{K} \cup L$ for a two-dimensional H-invariant set L. The developing map **dev** sends homeomorphic this set to a radiant half-space with a line in the boundary passing through O removed. Again the following lemma gives us a contradiction.

LEMMA C.10. *Let N be a radiant affine manifold homeomorphic to a torus times an interval with totally geodesic boundary. Let (D, j) be the development pair of N and $j(\pi_1(M))$ is in one of the above four cases D, P, U, or C. Let x, y, and z denote the standard coordinate functions of $\mathbf{R}^3$. Then a developing map of N can not be as follows: it maps $\tilde{M}$ into (but not necessarily onto) a half-space given by $z \geq 0$, and the two boundary components of $\tilde{M}$ respectively homeomorphic onto two components of $z = 0$ with the line given by $y = 0$ or $x = 0$ removed.*

PROOF. As N is homeomorphic to a torus times an interval, an ordered choice of two generators of G' induces orientations on each leaf torus. Given an orientation on N and the boundary orientation on ∂N, the generator orientation agrees with the boundary orientation at one component of ∂N and disagrees at the other component.

Look at $\tilde{\partial} N$ mapping homeomorphic to the two components A_1 and A_2 of the xy-plane with a line removed, say $x = 0$. Then since G acts on A_1 and A_2, the ordered choice of generators induces an orientation on each component. As $-\mathrm{I}$ is orientation-preserving on the xy-plane, and commutes with G, it follows that the generator orientation agrees with the boundary orientation of the half-space given by $z \geq 0$ or disagrees with it at both components A_1 and A_2. However, as $\mathbf{dev}(\tilde{N})$ lies in the half-space, the boundary orientations of A_1 and A_2 are from the boundary orientation of ∂N, a contradiction. □

In case D, let A and L be as above. The open set $\mathbf{dev}(A)$ is a cone with three adjacent two-dimensional H-invariant sets. In its boundary, $\mathbf{dev}(\tilde{K} \cup L)$ includes only two of the adjacent two-dimensional H-invariant sets. Let m be the remaining H-invariant set. Then there exists a coordinate function z on $\mathbf{R}^3$ such that $z = 0$ is a plane through O including m. As above, dz/z induces a closed 1-form on N_1 taking positive values under the radial vector field. Thus N_1 admits a total cross-section by Lemma C.3.

Similarly, we can show that each N_i admits a total cross-section to the radial flow. We can patch the cross-sections together at the tori, and obtain a total cross-section for M. The reason is that each N_i is affinely homeomorphic to each other by maps induced by reflections along two-dimensional H-invariant sets. Hence, we may assume that the homotopy classes of the total cross-sections are the same. This shows that M is a generalized affine suspension. There is a more detailed description of this case in section 3.2 of [**3**].

We now study the last but most complicated case P: Here, we are given standard coordinate functions x, y, and z so that our group H takes the form P.

As the xz-plane and the xy-plane are H-invariant, the interior of $\mathbf{dev}(\tilde{N}_1)$ may be given by $y > 0$ and $z > 0$ (up to sign changes of coordinate functions). The boundary parts $\mathbf{dev}(\tilde{K})$ and $\mathbf{dev}(L)$ are obtained from intersecting the planes given by $y = 0$ and $z = 0$ with $\mathbf{dev}(\tilde{N}_1)$. As G acts on these sets so that the quotients are tori, we may assume without loss of generality that $\mathbf{dev}(\tilde{K})$ is one of the following form: $y = 0, z > 0, x > 0$; $y = 0, z > 0, x < 0$; or $y > 0, z = 0$; and similarly, $\mathbf{dev}(L)$ is of form: $y > 0, z = 0$; $y = 0, z > 0, x > 0$; or $y = 0, z > 0, x < 0$.

As there are exactly two adjacent two-dimensional H-orbits, we have two cases:

(i) $\mathbf{dev}(\tilde{K})$ and $\mathbf{dev}(L)$ are both triangles given by $y = 0, z > 0, x > 0$ and $y = 0, z > 0, x < 0$ respectively.
(ii) If $\mathbf{dev}(\tilde{K})$ is the open lune given by $y > 0, z = 0$, then $\mathbf{dev}(L)$ must be a triangle given by $y = 0, z > 0, x > 0$ or $y = 0, z > 0, x < 0$ and vice versa.

We define a Euclidean reflection R about xy-plane. Then the group F generated by $-\mathrm{I}$ and R is of order four. As elements of H commute with F, we see that the closure in $\tilde{M}$ of each three-dimensional H-invariant set in $\tilde{M}$ under $\mathbf{dev}$ maps homeomorphic to $\mathbf{dev}(A) \cup \mathbf{dev}(\tilde{K}) \cup \mathbf{dev}(L)$ as in (i) or (ii) or an image regions of type (i) or (ii) under an element of F.

If every N_i is of type (i), we easily see that $\mathbf{dev}(\tilde{N}_i)$ lies in the set $z > 0$ by induction. Thus $\mathbf{dev}(\tilde{M})$ lies in the same set. Hence, the closed 1-form dz/z is G-invariant, and induces a closed 1-form on N_1 which takes a non-zero value under radiant vectors. This means that N_1 has a total cross-section by Lemma C.3. Hence M is a generalized affine suspension.

From now on we assume that each N_i is of type (ii). Thus, N_i and N_{i+1} meet at a torus which is a quotient of a triangle or a lune alternatively according to i.

Suppose that N_j and N_{j+1} meet at a torus which is a quotient of a triangle. Then we see that the remaining boundary components of $\tilde{N}_j \cup \tilde{N}_{j+1}$ map homeomorphic under $\mathbf{dev}$ to open lunes which are components of a plane with a line removed. This is a contradiction by Lemma C.10. Therefore M equals N_1 or the union of only two N_1 and N_2 meeting a torus which is a quotient of a lune. (We remark that this corresponds to a generalized affine suspension of π-annulus of type C.)

We will show that N_1 admits a total cross-section. Since N_2 can be obtained by the reflection R commuting with elements of G, this will show that M has a total cross-section.

We may assume without loss of generality by conjugations that the connected two-dimensional subgroup H of matrices of form P is of form:

$$\begin{pmatrix} e^a & be^a & 0 \\ 0 & e^a & 0 \\ 0 & 0 & e^c \end{pmatrix}$$

where a, b, c lies in a two-dimensional subspace of $\mathbf{R}^3$ with coordinate functions a, b, and c. A lattice L in P determines a subspace of $\mathbf{R}^3$, to be denoted by $P(L)$. Let the ac-plane have the orientation given by $\{e_a, e_c\}$ and the ab-plane have the orientation by $\{e_a, e_b\}$ where e_a, e_b, and e_c are unit vectors in the positive parts of the a-, b-, and c-axes respectively.

LEMMA C.11. *Let L be a lattice in a connected two-dimensional subgroup of group of matrices of form P. Let U be the domain given by $y > 0, z > 0$ union with the set U_{xz} given by $x > 0, y = 0, z > 0$ and the set U_{yz} given by $y > 0, z = 0$. Then L acts on U so that U/L is a manifold if and only if for the projections $p_{ac} : \mathbf{R}^3 \to \mathbf{R}^2$ to the ac-plane and $p_{ab} : \mathbf{R}^3 \to \mathbf{R}^2$ to the ab-plane, $g = p_{ac} \circ (p_{ab}|P(L))^{-1}$ is orientation-reversing.*

PROOF. Suppose that U/L is a manifold. Then U/L is homeomorphic to a torus times an interval. Let L' be a lattice in $P(L)$ corresponding to L by the above description. Then a connected two-dimensional group $\tilde{H}$ of elements of the above form with $a, b, c \in P(L)$ acts on U properly and without fixed points. We see that U is foliated by $\tilde{H}$-orbits. Thus, U/L is foliated by leaves that are $\tilde{H}$-orbits quotient out by L, homeomorphic to tori. Using an orientation on $P(L)$, each leaf has an induced orientation. We see easily that the leaf-space is homeomorphic to an interval and the boundary of U/L corresponds to the endpoints of the interval.

Consider the map $\mathcal{F} : \mathbf{R}^3 \to U^o$ given by

$$(a, b, c) \mapsto \begin{pmatrix} e^a & be^a & 0 \\ 0 & e^a & 0 \\ 0 & 0 & e^c \end{pmatrix} \begin{pmatrix} 1 \\ 1 \\ 1 \end{pmatrix} = ((b+1)e^a, e^a, e^c)$$

which is a homeomorphism. Also, define $\mathcal{F}_{ac} : \mathbf{R}^2_{ac} \to U_{xz}$, where $\mathbf{R}^2_{ab}$ denotes the ab-plane, by

$$(a, 0, c) \mapsto \begin{pmatrix} e^a & be^a & 0 \\ 0 & e^a & 0 \\ 0 & 0 & e^c \end{pmatrix} \begin{pmatrix} 1 \\ 0 \\ 1 \end{pmatrix} = (e^a, 0, e^c)$$

Define $\mathcal{F}_{ab} : \mathbf{R}^2_{ab} \to U_{xy}$, where $\mathbf{R}^2_{ab}$ denotes the xy-plane, by

$$(a, b, 0) \mapsto \begin{pmatrix} e^a & be^a & 0 \\ 0 & e^a & 0 \\ 0 & 0 & e^c \end{pmatrix} \begin{pmatrix} 1 \\ 1 \\ 0 \end{pmatrix} = ((b+1)e^a, e^a, 0).$$

Under $\mathcal{F}$, each $\tilde{H}$-orbit corresponds to a translate of $P(L)$. We may add $\mathbf{R}^2_{ac}$ and $\mathbf{R}^2_{ab}$ to $\mathbf{R}^3$ by considering $(a + t, e^{-t} - 1, c)$, $t < 0$, to converge to $(a, 0, c)$ as

$t \to -\infty$ and considering (a, b, t), $t < 0$, to converge to $(a, b, 0)$ as $t \to -\infty$. Then $\mathbf{R}^3$ is an open subspace of the completed space C. By adding $\mathcal{F}_{ac}$ and $\mathcal{F}_{ab}$ to $\mathcal{F}$, we obtain a homeomorphism $C \to U$ with appropriate topology on C.

In U^o, a $\tilde{H}$-orbit separates U_{xz} and U_{xy} since they correspond to two boundary components of a torus times an interval. In C, $P(L)$ must separate $\mathbf{R}^{ac}$ and $\mathbf{R}^{ab}$. Considering g to be a map $\mathbf{R}^2 \to \mathbf{R}^2$, as $g(a, b) = (a, b')$ for each (a, b) and some b', it follows that g is represented by a matrix

$$\begin{pmatrix} 1 & l \\ 0 & m \end{pmatrix}$$

where the plane $P(L)$ is given by $c = la + mb$.

Consider $\mathbf{R}^3$ as an open upper-hemisphere, i.e., as an affine patch, in the projective sphere $\mathbf{S}^3$ with boundary $\mathbf{S}^2$. Since the point that $(x+t, e^{-t}-1, z)$ converges on $\mathbf{S}^2$ as $t \to -\infty$ equals the ray through $(0, 1, 0, 0)$ and (x, y, t) converges to the ray through $(0, 0, -1, 0)$, for separation to hold, we must have that the function $z - lx - my$ takes different signs at these two points. This means that $m < 0$. Therefore the determinant of the matrix equals m which is negative.

We now prove the converse. Using the notation above, we see that $m < 0$. Let l_1 be the arc given by $\{(t, e^{-t} - 1, 0) | t \leq 0\}$ and l_2 one by $\{(0, 0, s) | s \leq 0\}$. Then $l_1 \cup l_2$ is an arc with well-defined endpoints at $\mathbf{R}^2_{ac}$ and $\mathbf{R}^2_{ab}$ in C. The arc α meets each of the translates of $P(L)$ at a unique point eventually far away. We can easily choose an arc α' which eventually agrees with α far away and meets each translate of $P(L)$ at a unique point. The image of α' by $\mathcal{F}$ is an arc in U with endpoints in U_{xz} and U_{xy} which meets each $\tilde{H}$-orbit at exactly one point. As L acts on each $\tilde{H}$-orbit to produced a torus, it follows that this condition is enough to give us a compact fundamental domain in U of the L-action. Hence, U/L is a compact manifold homeomorphic to a torus times an interval. □

REMARK C.1. There are analogous statements in case D also. But we omit them here.

Let L' be the lattice in $\mathbf{R}^3$ corresponding to our group G by above correspondence. Let ϑ be an element of L' so that $a - c$ and b values are positive on it. We can always choose such ϑ since $m < 0$ for our group L' by the above lemma C.11. We may further assume that ϑ is not a power of an element of L'. Let ϑ' be the corresponding element of G. The condition implies that for the projective automorphism ϑ'' corresponding to ϑ' acting on $\mathbf{S}^2$, $\langle \vartheta'' \rangle$ acts properly and freely on the subset U' of $\mathbf{S}^2$ corresponding to U under the radial projection as $[0, 0, 1]$ corresponds to a repelling fixed point of ϑ'' (see Section 1.4 of [**12**]). Hence, $U'/\langle \vartheta'' \rangle$ is a compact real projective annulus with geodesic boundary (of type *IIb* in [**12**]). Hence, $U/\langle \vartheta' \rangle$ is an $\mathbf{R}$-bundle over this quotient space $U'/\langle \vartheta'' \rangle$. We see easily that $U/\langle \vartheta' \rangle$ has a total cross-section A with real projective structure isomorphic to $U'/\langle \vartheta'' \rangle$. As the deck-transformation group is abelian of rank 2, the group also acts on $U/\langle \vartheta' \rangle$ properly discontinuously. There exists an element φ of G inducing an automorphism φ' on $U/\langle \vartheta' \rangle$ so that $(\varphi')^i(A)$ is disjoint from A for $i \neq 0$. Since $U'/\langle \vartheta, \varphi \rangle$ is a finite cover of N_1 with a total cross section coming from A, N_1 admits a total cross-section.

EXAMPLE C.1. Let ϑ be equal to

$$\begin{pmatrix} e^2 & e^2 & 0 \\ 0 & e^2 & 0 \\ 0 & 0 & e^{-4} \end{pmatrix}$$

and φ equal to

$$\begin{pmatrix} e^3 & -e^3 & 0 \\ 0 & e^3 & 0 \\ 0 & 0 & e^3 \end{pmatrix}.$$

Then the group $\langle \vartheta, \varphi \rangle$ acts on U freely and properly discontinuously. The above coefficients are given by $l = -1/5$ and $m = -18/5$. This has a total cross-section.

Finally, suppose that both types (i) or (ii) occur for our given manifold M, which decomposes into $N_1, \dots, N_n$ in a sequential manner. We note that type (ii) ones occur in adjacent pairs or adjacent to boundary components of M. Type (i) submanifolds can occur in sequence or alone adjacent to boundary components. Pick N_j which is of type (ii), and choose a generator ϑ as above. We can suppose that $\mathbf{dev}(\tilde{N}_j)$ is given by the region U in Lemma C.11. Then $[0, 0, 1]$ is a repelling fixed point of the projective automorphism ϑ' on $\mathbf{S}^2$ corresponding to ϑ.

Recall that A_i is the H-invariant domain given by $\mathbf{dev}(\tilde{N}_i)$ for $i = 1, \dots, n$, and the group F generated by a reflection R and $-\mathrm{I}$. Recall also that up to an action of F, A_i can be of the form the open set $y > 0$ and $z > 0$ union with

- the two sets in $z > 0$ and $y = 0$; i.e., the open disk D_1 given by $x > 0$ and D_2 given by $x < 0$ respectively.
- the open disk D_3 given by $y > 0$ and $z = 0$, and D_1.
- D_3, and D_2.

Clearly $\langle \vartheta' \rangle$ acts properly on the domains in $\mathbf{S}^2$ corresponding to first and second cases as it acts properly on U'. We claim that the third case is actually impossible to occur as $\mathbf{dev}(\tilde{N}_i)$ for some i. Since H acts properly and freely on $\tilde{N}_i$, $\tilde{N}_i$ is foliated by H-orbits which map to tori foliating N_i under the covering map. Let L be an H-orbit in A_i^o with A_i in the above forms. L/G is a torus in N_i separating two boundary components of N_i regardless of what form A_i is. As the first two cases are possible, L separates D_1 and D_2, and L separates D_1 and D_3. We can see that L cannot separate D_2 and D_3 as well using the $\mathbf{Z}_2$-intersection theory. But L corresponds to an H-orbit in each $\mathbf{dev}(\tilde{N}_i)$. (As a dual reasoning, one can consider H-orbits of a point mapped by F lifted to A_is.)

Since every $\mathbf{dev}(\tilde{N}_i)$ is of first two forms up to an action of F, it follows that $\langle \vartheta' \rangle$ acts properly on Q which is a union of domains corresponding to regions of the first two types up to elements of F.

This means as before that $\tilde{M}/\langle \vartheta \rangle$ is an $\mathbf{R}$-bundle over a compact surface $Q/\langle \vartheta' \rangle$ with radial lines forming the fibers. Choosing a section and finding another deck transformation φ which induces an automorphism φ' on $\tilde{M}/\langle \vartheta \rangle$ so that the images of the section under φ'^i are all disjoint from each other; $\tilde{M}/\langle \vartheta, \varphi \rangle$ is a radiant affine 3-manifold with a total cross-section covering M finitely. We obtain that M admits a total cross-section.

Bibliography

1. R. Abraham and J. Tromba, *Foundations of mechanics*, Addison-Wesley, Redwood City, 1978.
2. T. Barbot, *Structures affines radiales sur les variétés de Seifert*, preprint.
3. ______, *Variétés affines radiales de dimension trois*, preprint.
4. ______, *Structures affines radiales sur les 3-variétés à monodromie résoluble*, preprint, Universidade Federal Fluminense, 1997.
5. Y. Benoist, *Nilvariétés projectives*, Comment. Math. Helv. **69** (1994), 447–473.
6. ______, *Tores affines*, preprint, 1999.
7. J. P. Benzécri, *Variétés localement affines et projectives*, Bull. Soc. Math. France **88** (1960), 229–332.
8. M. Berger, *Geometry* I, Springer, New York, 1987.
9. Y. Carrière, *Autour de la conjecture de L. Markus sur les variétés affines*, Invent. Math. **95** (1989), 615–628.
10. ______, *Questions ouvertes sur les variétés affines*, Séminaire Gaston Darboux de Géométrie et Topologie Différentielle, 1991-1992(Montpellier), Univ. Montpellier II. Montpellier, 1993, pp. 69–72.
11. S. Choi, *Convex decompositions of real projective surfaces.* I: *π-annuli and convexity*, J. Differential Geom. **40** (1994), 165–208.
12. ______, *Convex decompositions of real projective surfaces.* II: *Admissible decompositions*, J. Differential Geom. **40** (1994), 239–283.
13. ______, *Convex decompositions of real projective surfaces.* III: *For closed and nonorientable surfaces*, J. Korean Math. Soc. **33** (1996), 1138–1171.
14. ______, *The universal cover of an affine three-manifold with holonomy of discompactedness two*, Geometry, Topology, and Physics (Berlin-New York) (B. Apanasov et. al., ed.), W. de Gruyter, 1997, pp. 107–118.
15. ______, *Convex and concave decomposition of manifolds with real projective structures*, Mem. Soc. Math. France **78** (1999), 1–106.
16. ______, *The universal cover of an affine three-manifold with holonomy of shrinkable dimension $\leq$ two*, Int. Journal of Mathematics **11** (2000), no. 3, ?–?
17. S. Choi and W. M. Goldman, *The classification of real projective structures on compact surfaces*, Bull. Amer. Math. Soc. **34** (1997), 161–171.
18. S. Dupont, *Solvariétés projectives de dimension 3*, Ph.D. thesis, Université de Paris 7, 1998.
19. D. Fried, *Affine 3-manifolds that fiber by circles*, preprint I.H.E.S, 1992.
20. D. Fried, W. Goldman, and M. Hirsch, *Affine manifolds with nilpotent holonomy*, Comment. Math. Helv. **56** (1981), 487–523.
21. W. Goldman, *Affine manifolds and projective geometry on surfaces*, senior thesis, Princeton University, 1977.
22. ______, a letter to W. Thurston, 1984.
23. ______, *Convex real projective structures on surfaces*, J. Differential Geom **31** (1990), 791–845.
24. V. Guillemin and A. Pollack, *Differential topology*, Prentice Hall, New Jersey, 1974.
25. G. Hector and U. Hirsch, *Geometry of foliations, part B*, Aspects of Math., Friedr. Vieweg & Sohn, Braunschweig, 1987, 2'nd edition.
26. J. Hempel, *3-manifolds*, Annals of Math. Studies, vol. 86, Princeton Univ. Press, 1976.
27. M. Hirsch and W. Thurston, *Foliated bundles, invariant measures and flat manifolds*, Ann. Math. **101** (1975), 369–390.
28. V. Kac and E. B. Vinberg, *Quasi-homogeneous cones*, Math. Notes **1** (1967), 231–235, Translated from Mat. Zametki **1** (1967), 347–354.

29. Y. Kamishima and S. Tan, *Deformation spaces on geometric structures*, Aspects of Low Dimensional Manifolds (Tokyo) (Y. Matsumoto and S. Morita, eds.), Advanced Studies in Pure Math., vol. 20, 1992, pp. 263–299.
30. H. Kim and H. Lee, *The Euler characteristic of a certain class of projectively flat manifolds*, Topology and its App. **40** (1991), 195–201.
31. ———, *The Euler characteristic of projectively flat manifolds with amenable fundamental groups*, Proc. Amer. Math. Soc. **118** (1993), 311–315.
32. R. Kirby and L. Siebenmann, *Foundational essays on topological manifolds, smoothings and triangulations*, Annals of Mathematical Studies, vol. 88, Princeton University Press, 1977.
33. S. Kobayashi, *Projectively invariant distances for affine and projective structures*, Differential Geometry (Warsaw), Banach Center Publication, vol. 12, Polish Scientific Publishers, 1984, pp. 127–152.
34. S. Kobayashi and T. Nagano, *On projective connections*, Journal of Mathematics and Mechanics **13** (1964), 215–235.
35. B. Kostant and D. Sullivan, *The Euler characteristic of an affine space form is zero*, Bull. Amer. Math. Soc. **81** (1975), 937–938.
36. J. L. Koszul, *Déformations des connexions localement plates*, Ann. Inst. Fourier (Grenoble) **18** (1968), 103–114.
37. N. H. Kuiper, *On compact conformally Euclidean spaces of dimension* > 2, Ann. Math. **52** (1950), 487–490.
38. T. Nagano and K. Yagi, *The affine structures on the real two torus.* I, Osaka J. Math. **11** (1974), 181–210.
39. J. Ratcliffe, *Foundations of hyperbolic manifolds*, GTM 149, Springer, New York, 1994.
40. P. Scott, *The geometries of 3-manifolds*, Bull. London Math. Soc. **15** (1983), 401–487.
41. J. Smillie, *Affine manifolds with diagonal holonomy* I, preprint, 1977.
42. ———, *Affinely flat manifolds*, Ph.D. thesis, University of Chicago, 1977.
43. ———, *An obstruction to the existence of affine structures*, Invent. Math. **64** (1981), 411–415.
44. D. Sullivan and W. Thurston, *Manifolds with canonical coordinate charts: some examples*, Enseign. Math **29** (1983), 15–25.
45. W. Thurston, *Three-dimensional geometry and topology, vol. 1*, Princeton Mathematical Series, vol. 35, Princeton University Press, 1997.
46. D. Tischler, *On fibering certain foliated manifolds over* S^1, Invent. Math. **36** (1970), 153–154.

Editorial Information

To be published in the *Memoirs*, a paper must be correct, new, nontrivial, and significant. Further, it must be well written and of interest to a substantial number of mathematicians. Piecemeal results, such as an inconclusive step toward an unproved major theorem or a minor variation on a known result, are in general not acceptable for publication. Papers appearing in *Memoirs* are generally longer than those appearing in *Transactions*, which shares the same editorial committee.

As of May 31, 2001, the backlog for this journal was approximately 7 volumes. This estimate is the result of dividing the number of manuscripts for this journal in the Providence office that have not yet gone to the printer on the above date by the average number of monographs per volume over the previous twelve months, reduced by the number of volumes published in four months (the time necessary for preparing a volume for the printer). (There are 6 volumes per year, each containing at least 4 numbers.)

A Consent to Publish and Copyright Agreement is required before a paper will be published in the *Memoirs*. After a paper is accepted for publication, the Providence office will send a Consent to Publish and Copyright Agreement to all authors of the paper. By submitting a paper to the *Memoirs*, authors certify that the results have not been submitted to nor are they under consideration for publication by another journal, conference proceedings, or similar publication.

Information for Authors

Memoirs are printed from camera copy fully prepared by the author. This means that the finished book will look exactly like the copy submitted.

The paper must contain a *descriptive title* and an *abstract* that summarizes the article in language suitable for workers in the general field (algebra, analysis, etc.). The *descriptive title* should be short, but informative; useless or vague phrases such as "some remarks about" or "concerning" should be avoided. The *abstract* should be at least one complete sentence, and at most 300 words. Included with the footnotes to the paper should be the 2000 *Mathematics Subject Classification* representing the primary and secondary subjects of the article. The classifications are accessible from `www.ams.org/msc/`. The list of classifications is also available in print starting with the 1999 annual index of *Mathematical Reviews*. The Mathematics Subject Classification footnote may be followed by a list of *key words and phrases* describing the subject matter of the article and taken from it. Journal abbreviations used in bibliographies are listed in the latest *Mathematical Reviews* annual index. The series abbreviations are also accessible from `www.ams.org/publications/`. To help in preparing and verifying references, the AMS offers MR Lookup, a Reference Tool for Linking, at `www.ams.org/mrlookup/`. When the manuscript is submitted, authors should supply the editor with electronic addresses if available. These will be printed after the postal address at the end of the article.

Electronically prepared manuscripts. The AMS encourages electronically prepared manuscripts, with a strong preference for $\mathcal{A}\mathcal{M}\mathcal{S}$-LaTeX. To this end, the Society has prepared $\mathcal{A}\mathcal{M}\mathcal{S}$-LaTeX author packages for each AMS publication. Author packages include instructions for preparing electronic manuscripts, the *AMS Author Handbook*, samples, and a style file that generates the particular design specifications of that publication series. Though $\mathcal{A}\mathcal{M}\mathcal{S}$-LaTeX is the highly preferred format of TeX, author packages are also available in $\mathcal{A}\mathcal{M}\mathcal{S}$-TeX.

Authors may retrieve an author package from e-MATH starting from `www.ams.org/tex/` or via FTP to `ftp.ams.org` (login as `anonymous`, enter username as password, and type `cd pub/author-info`). The *AMS Author Handbook* and the *Instruction Manual* are available in PDF format following the author packages link from `www.ams.org/tex/`. The author package can be obtained free of charge by sending email to `pub@ams.org` (Internet) or from the Publication Division, American Mathematical Society, P.O. Box 6248, Providence, RI 02940-6248. When requesting an author package, please specify $\mathcal{A}_\mathcal{M}\mathcal{S}$-LaTeX or $\mathcal{A}_\mathcal{M}\mathcal{S}$-TeX, Macintosh or IBM (3.5) format, and the publication in which your paper will appear. Please be sure to include your complete mailing address.

Sending electronic files. After acceptance, the source file(s) should be sent to the Providence office (this includes any TeX source file, any graphics files, and the DVI or PostScript file).

Before sending the source file, be sure you have proofread your paper carefully. The files you send must be the EXACT files used to generate the proof copy that was accepted for publication. For all publications, authors are required to send a printed copy of their paper, which exactly matches the copy approved for publication, along with any graphics that will appear in the paper.

TeX files may be submitted by email, FTP, or on diskette. The DVI file(s) and PostScript files should be submitted only by FTP or on diskette unless they are encoded properly to submit through email. (DVI files are binary and PostScript files tend to be very large.)

Electronically prepared manuscripts can be sent via email to `pub-submit@ams.org` (Internet). The subject line of the message should include the publication code to identify it as a Memoir. TeX source files, DVI files, and PostScript files can be transferred over the Internet by FTP to the Internet node `e-math.ams.org` (130.44.1.100).

Electronic graphics. Comprehensive instructions on preparing graphics are available at `www.ams.org/jourhtml/graphics.html`. A few of the major requirements are given here.

Submit files for graphics as EPS (Encapsulated PostScript) files. This includes graphics originated via a graphics application as well as scanned photographs or other computer-generated images. If this is not possible, TIFF files are acceptable as long as they can be opened in Adobe Photoshop or Illustrator. No matter what method was used to produce the graphic, it is necessary to provide a paper copy to the AMS.

Authors using graphics packages for the creation of electronic art should also avoid the use of any lines thinner than 0.5 points in width. Many graphics packages allow the user to specify a "hairline" for a very thin line. Hairlines often look acceptable when proofed on a typical laser printer. However, when produced on a high-resolution laser imagesetter, hairlines become nearly invisible and will be lost entirely in the final printing process.

Screens should be set to values between 15% and 85%. Screens which fall outside of this range are too light or too dark to print correctly. Variations of screens within a graphic should be no less than 10%.

Inquiries. Any inquiries concerning a paper that has been accepted for publication should be sent directly to the Electronic Prepress Department, American Mathematical Society, P. O. Box 6248, Providence, RI 02940-6248.

Editors

This journal is designed particularly for long research papers, normally at least 80 pages in length, and groups of cognate papers in pure and applied mathematics. Papers intended for publication in the *Memoirs* should be addressed to one of the following editors. In principle the Memoirs welcomes electronic submissions, and some of the editors, those whose names appear below with an asterisk (*), have indicated that they prefer them. However, editors reserve the right to request hard copies after papers have been submitted electronically. Authors are advised to make preliminary email inquiries to editors about whether they are likely to be able to handle submissions in a particular electronic form.

Algebra to CHARLES CURTIS, Department of Mathematics, University of Oregon, Eugene, OR 97403-1222 email: cwc@darkwing.uoregon.edu

Algebraic geometry and commutative algebra to LAWRENCE EIN, Department of Mathematics, University of Illinois, 851 S. Morgan (M/C 249), Chicago, IL 60607-7045; email: ein@uic.edu

Algebraic topology and cohomology of groups to STEWART PRIDDY, Department of Mathematics, Northwestern University, 2033 Sheridan Road, Evanston, IL 60208-2730; email: priddy@math.nwu.edu

Combinatorics and Lie theory to SERGEY FOMIN, Department of Mathematics, University of Michigan, Ann Arbor, Michigan 48109-1109; email: fomin@math.lsa.umich.edu

Complex analysis and complex geometry to DUONG H. PHONG, Department of Mathematics, Columbia University, 2990 Broadway, New York, NY 10027-0029; email: phong@math.columbia.edu

*__Differential geometry and global analysis__ to LISA C. JEFFREY, Department of Mathematics, University of Toronto, 100 St. George St., Toronto, ON Canada M5S 3G3; email: jeffrey@math.toronto.edu

*__Dynamical systems and ergodic theory__ to ROBERT F. WILLIAMS, Department of Mathematics, University of Texas, Austin, Texas 78712-1082; email: bob@math.utexas.edu

Functional analysis and operator algebras to BRUCE E. BLACKADAR, Department of Mathematics, University of Nevada, Reno, NV 89557; email: bruceb@math.unr.edu

Geometric topology, knot theory and hyperbolic geometry to ABIGAIL A. THOMPSON, Department of Mathematics, University of California, Davis, Davis, CA 95616-5224; email: thompson@math.ucdavis.edu

Harmonic analysis, representation theory, and Lie theory to ROBERT J. STANTON, Department of Mathematics, The Ohio State University, 231 West 18th Avenue, Columbus, OH 43210-1174; email: stanton@math.ohio-state.edu

*__Logic__ to THEODORE SLAMAN, Department of Mathematics, University of California, Berkeley, CA 94720-3840; email: slaman@math.berkeley.edu

Number theory to MICHAEL J. LARSEN, Department of Mathematics, Indiana University, Bloomington, IN 47405; email: larsen@math.indiana.edu

*__Ordinary differential equations, partial differential equations, and applied mathematics__ to PETER W. BATES, Department of Mathematics, Brigham Young University, 292 TMCB, Provo, UT 84602-1001; email: peter@math.byu.edu

*__Partial differential equations and applied mathematics__ to BARBARA LEE KEYFITZ, Department of Mathematics, University of Houston, 4800 Calhoun Road, Houston, TX 77204-3476; email: keyfitz@uh.edu

*__Probability and statistics__ to KRZYSZTOF BURDZY, Department of Mathematics, University of Washington, Box 354350, Seattle, Washington 98195-4350; email: burdzy@math.washington.edu

*__Real and harmonic analysis and geometric partial differential equations__ to WILLIAM BECKNER, Department of Mathematics, University of Texas, Austin, TX 78712-1082; email: beckner@math.utexas.edu

All other communications to the editors should be addressed to the Managing Editor, WILLIAM BECKNER, Department of Mathematics, University of Texas, Austin, TX 78712-1082; email: beckner@math.utexas.edu.

Selected Titles in This Series

(Continued from the front of this publication)

699 **Alexander Fel′shtyn,** Dynamical zeta functions, Nielsen theory and Reidemeister torsion, 2000

698 **Andrew R. Kustin,** Complexes associated to two vectors and a rectangular matrix, 2000

697 **Deguang Han and David R. Larson,** Frames, bases and group representations, 2000

696 **Donald J. Estep, Mats G. Larson, and Roy D. Williams,** Estimating the error of numerical solutions of systems of reaction-diffusion equations, 2000

695 **Vitaly Bergelson and Randall McCutcheon,** An ergodic IP polynomial Szemerédi theorem, 2000

694 **Alberto Bressan, Graziano Crasta, and Benedetto Piccoli,** Well-posedness of the Cauchy problem for $n \times n$ systems of conservation laws, 2000

693 **Doug Pickrell,** Invariant measures for unitary groups associated to Kac-Moody Lie algebras, 2000

692 **Mara D. Neusel,** Inverse invariant theory and Steenrod operations, 2000

691 **Bruce Hughes and Stratos Prassidis,** Control and relaxation over the circle, 2000

690 **Robert Rumely, Chi Fong Lau, and Robert Varley,** Existence of the sectional capacity, 2000

689 **M. A. Dickmann and F. Miraglia,** Special groups: Boolean-theoretic methods in the theory of quadratic forms, 2000

688 **Piotr Hajłasz and Pekka Koskela,** Sobolev met Poincaré, 2000

687 **Guy David and Stephen Semmes,** Uniform rectifiability and quasiminimizing sets of arbitrary codimension, 2000

686 **L. Gaunce Lewis, Jr.,** Splitting theorems for certain equivariant spectra, 2000

685 **Jean-Luc Joly, Guy Metivier, and Jeffrey Rauch,** Caustics for dissipative semilinear oscillations, 2000

684 **Harvey I. Blau, Bangteng Xu, Z. Arad, E. Fisman, V. Miloslavsky, and M. Muzychuk,** Homogeneous integral table algebras of degree three: A trilogy, 2000

683 **Serge Bouc,** Non-additive exact functors and tensor induction for Mackey functors, 2000

682 **Martin Majewski,** ational homotopical models and uniqueness, 2000

681 **David P. Blecher, Paul S. Muhly, and Vern I. Paulsen,** Categories of operator modules (Morita equivalence and projective modules, 2000

680 **Joachim Zacharias,** Continuous tensor products and Arveson's spectral C^*-algebras, 2000

679 **Y. A. Abramovich and A. K. Kitover,** Inverses of disjointness preserving operators, 2000

678 **Wilhelm Stannat,** The theory of generalized Dirichlet forms and its applications in analysis and stochastics, 1999

677 **Volodymyr V. Lyubashenko,** Squared Hopf algebras, 1999

676 **S. Strelitz,** Asymptotics for solutions of linear differential equations having turning points with applications, 1999

675 **Michael B. Marcus and Jay Rosen,** Renormalized self-intersection local times and Wick power chaos processes, 1999

674 **R. Lawther and D. M. Testerman,** A_1 subgroups of exceptional algebraic groups, 1999

673 **John Lott,** Diffeomorphisms and noncommutative analytic torsion, 1999

672 **Yael Karshon,** Periodic Hamiltonian flows on four dimensional manifolds, 1999

671 **Andrzej Rosłanowski and Saharon Shelah,** Norms on possibilities I: Forcing with trees and creatures, 1999